Bay Cities and Water Politics

DEVELOPMENT OF WESTERN RESOURCES

The Development of Western Resources is an interdisciplinary series focusing on the use and misuse of resources in the American West. Written for a broad readership of humanists, social scientists, and resource specialists, the books in this series emphasize both historical and contemporary perspectives as they explore the interplay between resource exploitation and economic, social, and political experiences.

Bay Cities and Water Politics

The Battle for Resources in Boston and Oakland

Sarah S. Elkind

Maps by Aaron J. Weier

Ellen —
A sample of what I DO when I know what I'm trying to do.
Thanks for everything.

Published by the University Press of Kansas (Lawrence, Kansas 66049), which was organized by the Kansas Board of Regents and is operated and funded by Emporia State University, Fort Hays State University, Kansas State University, Pittsburg State University, the University of Kansas, and Wichita State University

Library of Congress Cataloging in Publication Data

Elkind, Sarah S., 1963–
Bay cities and water politics : the battle for resources in Boston and Oakland / Sarah S. Elkind ; maps by Aaron J. Weier.
p. cm. — (Development of western resources)
Includes bibliographical references and index.
ISBN 0-7006-0907-5 (cloth : alk. paper)
1. Municipal water supply—Massachusetts—Boston. 2. Muncipal water supply—California—Oakland. 3. Water quality management. 4. Urban ecology—United States. 5. Municipal services- -Massachusetts—Boston. 6. Municipal services—California—Oakland. I. Title. II. Series.
HD4464.B67E45 1998
363.6'1'0974461—dc21 98-17798

British Library Cataloguing in Publication Data is available.

Printed in the United States of America

10 9 8 7 6 5 4 3 2 1

The paper used in this publication meets the minimum requirements of the American National Standard for Permanence of Paper for Printed Library Materials Z39.48-1984.

Contents

Acknowledgments

This book is dedicated to my parents, who along with many other teachers, inspired and encouraged me. They not only made it possible for me to pursue my interests, but showed me that I should settle for nothing less. I trace my initial fascination with environmental history to my mother's passion for environmental education and the way my parents together made outdoor adventures and the history of our destinations so thoroughly a part of my childhood.

I take great pleasure in acknowledging the unflagging support extended to me by Martin V. Melosi, Donald Pisani, and Joel A. Tarr. They welcomed me into the discipline and gave freely of their time and their enthusiasm when I needed it most. Their comments and questions helped me refine my thinking at a number of crucial stages. All graduate students and young scholars should have such mentors.

This research would not have been possible without the assistance of many archivists and librarians. In Massachusetts, the staffs of the State Library, Boston Public Library, Massachusetts State Archives, the Medford, West Boylston, and Clinton Public Libraries, and the Boylston Historical Society were enormously helpful. Mary Lydon at the Massachusetts Water Resources Authority was also generous. Captain Al Swanson and Sean Fisher shared their prodigious knowledge of the Metropolitan District Commission archives and photographic collections.

In California, the archivists of the Bancroft Library and Water Resources Center Archives made trips to Berkeley a joy. The staffs of the Oakland History Room at the Oakland Public Library, the California State Archives, and the Richmond, Lodi, and Stockton Public Libraries were also very helpful. No one in California did more to make this project possible than Leo Bald and the Records Office staff at the East Bay Municipal Utility District. They have been extremely gracious whenever I have sought them out, even when I moved into their office for two months in 1991.

Martin Pernick, Richard Tucker, and Andrew Achenbaum at the University of Michigan were steadfast in their support. Donald Swann shared his enthusiasm for harbor history while we were both working in the

Boston Harbor Islands State Park. Our expeditions to examine old maps at the Boston Public Library eventually led me here.

The Urban Harbors Institute in Boston, Massachusetts and the Sourrisseau Academy of San Jose, California provided support for my dissertation research. The University of Wisconsin University Personnel Development Committee contributed research and travel funds as I revised the manuscript.

I would also like to thank Hal Rothman, Nancy Scott, and the staff of University Press of Kansas. Eugene Moehring and Harold Platt, who read the manuscript for the press, gave me many useful suggestions.

I have incurred many other debts for support, friendship, and hospitality during my endless travels to research and write this book. There are too many people to mention by name, but I do want to single out Peggy Shaffer, Nancy Jacobs, Bill Cronon, my fellow University of Michigan graduate students, and Octo and Sarah Barnett.

Finally, I owe much of the credit for the final form of this manuscript to Susan Hartzell. She edited the draft with extraordinary attention, insight, and caring, and made this a far better work than it would have been.

Introduction

Working in the Boston Harbor Islands State Park in the mid-1980s, I could ignore neither harbor pollution nor the unfolding drama of the Boston Harbor Cleanup. By 1985, a lawsuit had stripped one of the nation's oldest regional urban agencies, the Metropolitan District Commission, of its responsibilities over water delivery and waste disposal. The Metropolitan District Commission lost public faith so dramatically for a variety of reasons. The agency had grown increasingly isolated from public opinion. Political appointees in key positions did not always bring public works expertise to their jobs, and they owed greater allegiance to the state governor than to water and sewer customers. Agency employees tailored their actions to the needs of the institution and had few means to respond to public debate or ecological change. Without political or institutional incentives to respond to voters' desires, the Metropolitan District Commission could and did ignore changing public priorities for the harbor.

Of course, the problems that eventually brought down the Metropolitan District Commission were not entirely of a political nature. By the 1980s, Boston confronted some serious environmental problems that no single agency could have prevented. The city's hundred-year-old public works infrastructure had literally set water and sewer policies in stone; realistically, the city could not suddenly change the way its residents used water or disposed of its wastes. Decades of urban growth increased direct environmental pressures on the harbor. At the same time, conflicting demands for use of harbor resources grew ever more problematic, as waste disposal crowded out the shellfishing and finfishing industries and interfered with recreation. By the 1980s, the Metropolitan District Commission's inability to resolve these conflicting resource demands not only landed the agency in court, but also cost taxpayers billions for new sewage treatment facilities, and discredited Michael Dukakis' presidential campaign.[1]

What critics of the Metropolitan District Commission did not realize in the 1980s was that the characteristics of the agency that they blamed for harbor pollution, including its political isolation and narrow policy objectives, were originally intended to produce the major benefits of a metropolitan public works administration. From the 1880s on, many cities saw metropolitan or regional administration as an ideal means to improve the efficiency of public services and to reduce resource competition among urban neighbors. Political isolation and authority over only a few services seemed the best way to ensure that agencies implemented policies based on scientific principles rather than on political horse-trading. Over time, the very efficiency of these water and sewer systems permitted cities to grow so much that they overwhelmed their nineteenth- and early-twentieth-century service networks. Moreover, decreased voter influence over public bureaucracies prevented regional agencies from responding in a timely fashion to shifting public priorities. The fact that regionalism is now seen as a cause of contemporary urban environmental problems is an ironic consequence of late-nineteenth- and early-twentieth-century urban ambitions.

This study examines the origins of the urban regional institutions responsible for contemporary problems like the pollution of Boston Harbor. The comparison of regional public works in Boston, Massachusetts, and the East Bay cities of California explores the way physical conditions and public reactions to them prompted transformations in urban services, in the concept of public responsibility, and in the relationships among cities and their suburbs and rural hinterlands. The choice of these two urban centers is of course somewhat arbitrary. Boston merits examination not merely because its regional agencies came under fire in the 1980s, but because this was the first city to apply the old special district form of government to multiple-city, multiple-county administration of public works. East Bay regionalism, which followed three decades later, provides insight into how the metropolitan district became fully incorporated into the vocabulary of American political reform.

The evolution of metropolitan services in these two communities reflected Boston's and the East Bay's physical characteristics as well as their polictical histories. Their natural environments influenced how urban residents understood regionalism and public works. Boston's drainage problems directed attention to local river systems and the way the movement of drinking and waste water bound communities together. In contrast, water scarcity reinforced the East Bay's concept of water as an abstract commodity to be traded, divided, and manipulated by the marketplace and government agencies. The fact that regional agencies themselves served similar purposes in metropolises so distinct from one another emphasizes the unique role of the special district in urban politics and development.

Regionalism—the creation of metropolitan special districts—acknowledged the reconception of cities' relationships to the natural world in a number of ways.[2] The transfer of responsibility for water supply and sewerage from the individual to the public and from small to ever larger physical and governmental structures illustrates the expanding awareness of interconnection, first between one household and the next, then between neighborhoods, cities, and watersheds, and finally between adjacent drainage basins, bays, and states. Even as centralization decreased the resource competition that had divided metropolitan areas, regional agencies' abilities to claim and develop resources for urban residents forced state officials to arbitrate among communities and economic sectors all competing for the same resources. Although urbanites would not apply the label "environmental" to these resource problems until the late twentieth century, the evolution of regional networks forced them to acknowledge the connections between their communities and much larger, complex human and natural systems.

Insofar as regionalism forced Americans to confront resource distribution policies, metropolitan public works were one component of much broader debates over the best uses of natural resources taking place between the 1870s and 1930s. In these controversies, sewage disposal was as much a question of resource control as was diverting water for other domestic or industrial uses. Because household and factory wastes deposited in rivers precluded other water uses, the power to pollute a river or bay was no less the power to use water than was the right to divert a river into a field or factory. Sewerage and water supply are deeply connected; some of the nation's earliest water regulations grew out of efforts to protect water supplies from sewage contamination. The extension of urban water supply networks into distant rural communities often prevented those communities from building new sewers or new water systems. The water rights implicit in Boston's and the East Bay's sewerage systems, with their ocean waste outfalls, may appear less clear. However, these systems in effect reserved their harbors for waste disposal rather than for fishing, recreation, or wildlife habitat.

Enthusiasm for regionalism was not confined to Boston and the East Bay. Between the 1880s and the 1930s, metropolitan special districts were created throughout the country to provide a wide range of services. Many communities, Boston and the East Bay included, adopted regional solutions to specific environmental crises including water shortages and sewage pollution. Such crises fostered constituent demand for improved services and therefore stimulated both government growth and political reform.[3] Frequently, proposals for new public works grew from a discernible decline in service, as when protracted drought failed to replenish East Bay reservoirs year after year. Nevertheless, such direct relationships cannot always be found between environmental changes and public activism. Moreover, at many stages in the evolution of regional public works, residents of Boston and the East Bay rather suddenly responded to conditions which they had tolerated for many years. In my examination of the political implications of such environmental conditions, I have accepted contemporary assessments without endeavoring to distinguish between "real" and "perceived" crises.

Not every regional special district can be explained as a response to crisis. In many parts of the country, medium-sized cities with metropolitan ambitions built regional networks before their private or municipal services failed. In these cases, boosters promoted regionalism to attract business or to compete with rival communities.[4] Technology, too, played an important role in the trend toward regionalism. As innovation in pipe production, water appliances, and filtration systems made complex waterworks more affordable, voters began to demand the application of these new technologies.[5] Consulting engineers, hired to study municipal water and sewerage, frequently endorsed specific administrative as well as technological innovations that they had seen succeed in other communities. Further, state-imposed limits on municipal debt and taxation created a greater need for semiautonomous institutions with access to unrestricted sources of revenue.[6] Finally, special districts proliferated because they could successfully coordinate the management of extensive, complex service networks. Regionalism was, therefore, well adapted to the growing ecological understanding of the urban community as part of larger natural and human systems.

Regionalism as Politics

Building regional special districts required city residents and their leaders to increase public authority over urban services and natural resources. This authority evolved gradually in the United States. In the earliest efforts to improve water supplies and sanitation, individuals dug household wells and laid their own drainage pipes. When private efforts no longer sufficed, they turned to local officials or to private utility companies. Then, when the resulting municipal or corporate enterprises also proved inadequate, as they did in so many cities between 1880 and 1920, multiple-city and finally regional projects gained popular support. The history of urban water supply and sewerage, therefore, illuminates the relationship between public services and government growth in American cities. Moreover, a study of the evolution of regional public works demonstrates how the quest for resources encouraged the physical extension of urban power in an age characterized by great ambivalence toward cities.

Boston was a pioneer in public administration of urban services. There, city services and municipal claims on natural resources evolved as an outgrowth of government responsibility for public health. Boston's city officials took over local sewerage in 1823 and water supply in 1848. In 1889, the creation of the Metropolitan Sewerage Commission made Boston the first American city to turn public services over to a metropolitan special district. With their approval of the Metropolitan Water Board in 1895, Bostonians signaled their wholehearted endorsement of an administrative form that had previously been used primarily to coordinate irrigation, education, and transportation infrastructure in rural areas.[7]

Public services emerged and expanded so early in Boston because of its geographical characteristics and the traditions that governed water resource distribution in Massachusetts. Until the early 1800s, permeable soils provided Bostonians with plentiful springs for local use. But as more people moved to the Shawmut peninsula, this same soil permeability led to problems. Wastes saturated the soil, contaminating springs and impeding drainage. When municipal waterworks replaced wells and cisterns in midcentury, water use increased dramatically; this compounded drainage and pollution problems. At the same time, urban growth increased pollution in all Boston area rivers, further aggravating Boston's sanitation and water supply problems. By the 1870s, upstream pollution forced Boston officials to exert their authority outside city limits. The suburbs, while resenting Boston's heavy-handedness, began to recognize their own public services as inadequate. As a result, regionalism gained popularity as a solution to the environmental and service crises, the resource competition, and the political conflicts that divided the greater Boston area.

Although political conflicts inside Boston and among rival communities made regionalism controversial, the discussion of public works development in Boston focused on the connections between environmental conditions and public health. By the 1850s, medical and public health professionals had come to associate dirt with disease; furthermore, they proscribed filthy living conditions as a major source of moral decay. Eventually, urban reformers extended the hazards of urban filth to include political as well as moral and physical corruption. By the end of the nineteenth century, these moral-environmental theories dominated public health and social reform ideology in the United States and Europe.

In Boston, moral-environmental theories directed municipal attention and public apprehension to the contamination of wells and reservoirs and the concentration of domestic sewerage along the waterfront. Local leaders saw improved water and sewer services in general, and regionalism in particular, as crucial to their efforts to combat disease by physically cleaning the city. Residents endorsed municipal and regional projects not only because they feared disease, but also because they feared the social chaos that the moral-environmental theorists found in filthy, impoverished neighborhoods. Significantly, the way in which regionalism diluted municipal officials' and voters' control over local policies rarely entered discussions of Boston's regional networks. The urgency of sanitary reform eclipsed these political concerns and ensured widespread support for centralized administration of urban public works even in cities as politically divided along partisan and ethnic lines as Boston.[8]

In 1923, when Oakland, Berkeley, and Richmond united to create the East Bay Municipal Utility District, Boston's metropolitan agencies served as a model for public waterworks administration. As in Boston, East Bay regionalism was celebrated as a solution to a wide range of local urban problems. These similarities are all the more striking given the differences between the cities' public services before regionalism. By the 1920s, germ theory had replaced moral-environmental explanations of disease. Without public health to galvanize support for municipal waterworks and sewerage, public authority over services and the urban environment was significantly weaker in the East Bay than in Boston. Meanwhile, water scarcity made aggressive water development both essential for urban growth and extremely attractive to entrepreneurs. Together, these factors created a powerful opposition to public works which delayed government-sponsored water development for decades.

In Oakland, household wells began to give out in the 1860s, only a few years after municipal incorporation. So, long before local residents were willing to tax themselves or authorize public waterworks, East Bay homeowners found themselves in need of fairly elaborate water systems. They looked to private enterprise to dam streams, dig new wells, and lay water pipes to their doors. These private water companies lasted for sixty years and became a major political force in the region.

Sewerage was a different story. By the 1850s, when East Bay residents approved their first municipal charters, most urbanites did not question that public agencies should provide drainage and other non-revenue-producing services. The comparison with Boston is telling. In Massachusetts, because neither water nor sewers made for good entrepreneurial investment, Boston officials assumed responsibility for both these services. In Oakland, however, public authority was extended only to those services, like sewers, that did not appeal to investors.

Because of California's tradition of limited public authority and because they could not use a public health crisis to generate support for their proposals, East Bay reformers had to base their public works campaigns on other issues. Eventually, they sold public water development as a panacea for a host of political and economic concerns. By the Progressive Era, Californians had come to blame the state's political, social, and economic woes on railroad corporations.[9] Over time, these accusations had rubbed off onto many other utility companies, including those that supplied water to urban communities. The East Bay water companies resembled the railroads in their monopoly of vital services, their heavy-handed manipulations of local politics and their resistance to nearly all regulatory efforts. Although this history gave anti-utility rhetoric an almost universal appeal, the East Bay water companies vigorously resisted public ownership. Bolstered by their monopoly of water resources, East Bay water companies defeated municipal and multiple-city waterworks initiatives for decades before finally succumbing to regionalism. Significantly, by the time East Bay voters finally approved the East Bay Municipal Utility District, the remaining private water company had neither enough water for its customers nor any practical means to increase supplies. Only the potent combination of water shortages, feared because of the dire consequences for continued economic growth, and antipathy toward private utilities could overcome East Bay residents' resistance to increased government authority.

In addition to increasing public authority, Boston and East Bay metropolitan special districts created new bases of power for bureaucrats and technical experts who owed little to municipal leaders or party organizations. Formation of the districts significantly reduced voters' oversight over public services in their communities and decreased the importance of elected officials in the management of, construction of, and hiring for local public works. For these reasons, one might expect to find municipal officials fighting regionalism. In fact, this was not the case. City leaders usually embraced the new agencies as a way to improve services without increasing taxes or exceeding state-imposed financial restrictions. They recognized that many of their constituents would welcome employment on public works projects. Furthermore, unlike many other Gilded Age and Progressive Era political reforms, special districts offered the benefits of multiple-city networks without threatening the political independence of central cities or their suburbs. Regionalism attracted widespread support because it offered the benefits of central, consolidated administration while preserving home rule.

Regionalism offered fewer benefits to the rural communities that surrendered their water or land for urban expansion. Despite some efforts to compensate the source regions, urban appropriation of rural resources aroused considerable resentment. The permanent presence of regional agencies in source communities continually reminded rural residents of what they had lost. Ethnic and class biases contributed to this rancor. As contractors brought hundreds of immigrant workers to rural communities to build massive public works, regional agencies came to symbolize the growing power of immigrants. Moreover, the appropriation of local resources violated the sense of ownership with which communities frequently regarded nearby landscapes, rivers, and ponds. These psychological irritants exacerbated the impact of the legal restrictions on water and land use imposed to protect urban projects, to say nothing of the wholesale destruction of homes, farms, factories, and towns that accompanied many reservoir projects. Thus, regionalism increased the tensions between urban and rural communities and frequently left source towns embittered and angry.

Unfortunately for source communities, rural opponents had few effective means to combat the concentration of urban political power that coincided with regionalism. New agencies granted cities unprecedented power outside their territorial boundaries and correspondingly reduced local control in source regions. Ironically, this centralization of power sapped the political as well as economic vitality of the very small towns that had embodied the ideals of American political and social community. In the end, regionalism permitted extensive urban political reform, but at the expense of municipal democracy as well as political and environmental self-determination in source regions. Despite rural overrepresentation in state legislatures, this transformation took place because of the environmental crises and powerful interests that supported improved urban public works. Among these interests, in both the East and the West, were advocates of industrial and urban expansion.

In Boston and East Bay regionalism political reforms were linked directly to the natural environment. Physical conditions, including urban pollution and resource shortages, played a crucial role in marshalling public support behind expensive and elaborate public works. The projects themselves expanded government authority by introducing new administrative entities and bringing ever larger physical territories under urban control. Political reformers hoped that regionalism would help combat urban political corruption, boosters anticipated an economic boom, and reformers seeking improved sanitation embraced regionalism's potential to transform urban society. The story of Boston and East Bay regionalism illustrates the ways in which urban environmental problems prompted political reform. It also sheds new light on both the expansion of government authority and the relationships of cities to natural resources and the environment in late-nineteenth- and early-twentieth-century America.

1 | Municipal and Private Services in the Nineteenth Century

Although many American metropolises eventually embraced regionalism, initially their residents did not see water supply, drainage, or other service needs in regional or even communal terms. Before municipal incorporation, individuals in most American towns built their own water supplies and arranged for their own waste disposal before they approved tax-funded efforts to assist with basic services. In many communities, municipal incorporation helped residents replace some of their private, individual structures with public, networked systems. Indeed, the desire for improved services often provided a major incentive for voters to approve municipal status for their communities. In other towns, municipal incorporation changed services relatively little because residents hesitated to expand public powers too broadly. Even in these communities, however, the wells and privies that were the backbone of household water supply and waste disposal could not sustain urban growth indefinitely. Private utility companies provided the services that local government could not, and they functioned much as municipal networks did. These early systems, whether held by public or private entities, drew attention to urban resource needs and laid the foundation for subsequent expansion of urban networks.

A number of factors determined whether cities relied on public or private service networks. Where communities had strong traditions of public activism, where residents could see the association between services and broad public mandates such as protecting public health, or where local conditions ruled out profitable private service development, citizens called upon their elected leaders to solve their common problems. Boston met all of these conditions by the time the city incorporated in 1823. Although Oakland, the largest of the East Bay cities, incorporated only thirty years after Boston, it never embraced public enterprise as wholeheartedly as its Massachusetts counterpart. Oakland voters accepted public sewer construction, but they refused to permit municipal waterworks. Water scarcity, the isolation of water issues from clear public responsibilities, and the profitability of private water services delayed public ownership of the water system for many decades. The differences between Boston's and Oakland's waterworks and sewers would have enormous impact on their respective efforts to create regional systems. These early networks established the nature of public authority in each community, even as they created the expectations for water and sewer systems that would eventually drive voters to abandon municipal for regional service administration.

Boston: The Public Service Tradition, 1823–1850

Throughout the nineteenth century, Boston's tradition of public activism facilitated municipal efforts to control the urban environment. Even before municipal incorporation, public officials had enormous discretion over both resources and services. Massachusetts common law reserved large ponds and lakes for public use. Town selectmen had authority to regulate private waste disposal and drainage and to sponsor the construction of public reservoirs for fire fighting. These precedents engendered an expansive sense of public responsibility, which served Boston well when new theories of disease and public demand for services began to place ever more emphasis on public remedies to urban environmental problems.

In the eighteenth and early nineteenth centuries, Boston residents developed elaborate but improvised household-based water supply and waste disposal structures. Although a few residents of the southern and western sections of the city purchased water from a private water company, for most Bostonians the water system consisted of hundreds of small, private wells and cisterns. Boston's sanitation was similarly decentralized. Nearly all the city's domestic wastes made their way into backyard privy vaults; these frequently overflowed. By the 1820s, a hundred years of private construction had left the city with a web of small sewers intended to drain only rain and standing water from private property. The shortcomings of decentralized water supplies and piecemeal sewerage were only too obvious. Wastes from privy vaults saturated Boston's soils and leached into private wells, fouling the water and spreading disease. Soot and cinders contaminated cisterns, rendering rainwater as unpalatable as groundwater. Fortunately, the traditions of public enterprise permitted government action. The public accepted the idea that such squalor, and the diseases and social problems associated with it, justified expanded municipal authority over utilities in Boston.

As they found their individual water and sewer structures lacking, Bostonians could not just turn over these responsibilities to public officials because the town government did not have authority to build elaborate public works. Furthermore, town leaders found even legal enterprises limited by Suffolk County's Court of Quarter Sessions, a body appointed by the governor to oversee town affairs. Municipal incorporation, however, would give the community freedom from state supervision as well as the power to expand public enterprise. The need for improved services figured prominently in the successful incorporation campaign. The 1823 charter granted the city authority over a number of public works, including sewers, fires, and street construction. City incorporation represented a significant change in the nature of local government. Boston voters, whatever their interest in public services, did not make the change lightly. In fact, they had rejected four earlier proposals for municipal incorporation in 1784, 1792, 1804, and 1815. The last lost by a mere thirty-one votes. Even after incorporation, Bostonians did not immediately embrace municipal activism. As the new city council gained the trust of voters, however, the city rapidly expanded the range of services it would provide. The transfer of sewer construction from the street department to a specialized sewer department in 1837 and the approval of a municipal water system in 1846 represented major steps in the expansion of Boston's municipal power.[1]

Water and sewer services fell to public officials because the nineteenth-century understanding of epidemics led many urban residents to perceive poor sanitation as a direct threat to their health. Before the 1830s or 1840s, most municipal health policies reflected the assumption that disease was caused by aberrant or careless individual behavior. Although cities empowered temporary health boards to enforce quarantines and isolate the sick, these efforts had only limited benefits. Moreover, they angered both the poor and the merchant classes. The poor grew increasingly resistant when city officials came to take the sick to pest houses. Meanwhile, merchants chafed under quarantines that closed ports and halted trade for months at a time. By substituting sanitary engineering for these older strategies, cities made a commitment to trade and to maintaining peace in poorer neighborhoods.[2]

By the middle of the nineteenth century, changes in theories of disease facilitated the expansion of public responsibility for water and sanitation. These theories rested on the association between dirt and disease as articulated by John Snow, Edwin Chadwick, and many others. In his 1842 sanitary survey of the English working class, Chadwick clearly articulated a moral-environmental theory central to nineteenth-century public works. According to Chadwick, the poor lived in misery because of the combined effects of disease, dirt, and vice. By attacking dirt, the easiest of the three to address through public action, he suggested, cities could protect all their residents.[3] The emphasis on dirt as a cause of disease placed health outside the responsibility of individuals and even made dirt in a city's poorest neighborhoods a threat to the health of the most influential members of society.[4] Popular authors like Victor Hugo brought these ideas to the public. Hugo portrayed Parisian sewers as a metaphor for the evils of urban society in his widely read novel, *Les Miserables.*[5] As one Boston editor noted, "It is impossible to make everybody cleanly [*sic*] . . . but it *is* possible to drain the whole city well and to canopy it above and to surround every dwelling with a tolerably pure and wholesome atmosphere."[6] Adequate sanitation seemed to promise a structural remedy for social problems. This theory of sanitation and social reform opened the way for greater public investment to control disease and directed ever more attention to the physical environment of the city.

Following Chadwick, many other social reformers latched onto this approach. For example, an article in the *Edinburgh Review* in 1849 justified the costs of municipal projects as investments that could ultimately decrease poor relief and prison expenses by rooting out the criminals who favored filthy districts. The authors argued here not only that conditions in the slums peeled away individuals' respect for the rules that governed society, but that urban squalor actually attracted criminals. They excused any invasion of privacy to "free the citizen from the vile fetters with which the acts of others have actually bound him, and to leave him free to pursue the natural tendency towards civilization and refinement."[7] The association of dirt, disease, and vice evolved into a social theory that touted the curative powers of clean and orderly physical surroundings for moral as well as physical ills. Class and ethnicity clearly influenced both Chadwick's research and the work of those who sought to implement his conclusions. Cleanliness became part of a complete program of social reform directed as much at the unfamiliar habits of the poor as at real vice and squalor. Safe water and adequate sewers, which were essential for cleanliness, therefore developed into a matter of social engineering, not merely of domestic convenience. Throughout the nineteenth century, the politics of public works such as water supply and sewerage expanded to reflect a host of political and social reform goals.

Boston's authority over sanitation and water supply was an extension of its historic power to pass laws related to the "causes of sickness, nuisances, and sources of filth that may be injurious to the health."[8] From the seventeenth century, Boston officials had exercised their authority to restrict the disposal of offal and animal remains in the Common, to eliminate privy vault nuisances and to encourage durable construction of private drains. As epidemics became more closely associated with filth, public health policies turned increasingly to sanitation, including both sewage and water supply as well as street cleaning, solid waste disposal, and school inspection. Bostonians accepted increased public responsibility for sanitation because they had begun to fear the contagious potential of unsanitary conditions not only next door, but in distant corners of the city. Increased public responsibility proved popular, in part, because the city could take on problems that individuals could not solve. Among reformers, of course, these problems included the complex social and political tensions so prominent in cities during this period. Boston established its precedents for public services early. These grew from the voters' desire for the protection of public access to resources, from high demand for urban services, and from the fear of unsanitary conditions that dominated public works thinking throughout the nineteenth century. All of these factors contributed to Bostonians' willingness to see sanitation as a public rather than a private responsibility.

Municipal Sewers

By the time of incorporation, Boston had outgrown water supply and waste disposal practices better suited to rural communities. Waste disposal problems appeared quickly as rubbish and sewage accumulated in Boston's streets and waterways. Residents responded to deteriorating conditions with demands that local officials intervene; ultimately, the promise of improved sewerage and other services convinced many Bostonians to support municipal incorporation. Boston's first city charter, approved in 1823, not only gave elected officials authority to build new drainage pipes, but also transferred ownership of all existing sewers to the new municipal institutions.

Government control of Boston's sewerage began with the regulation of private sewer construction and privy vault maintenance that actually predated municipal incorporation. Sewer regulations dating from the early eighteenth century specified construction techniques and the division of construction costs among neighbors. Because Bostonians built their storm drains independent of any oversight or planning and because construction specifications were rarely enforced, by 1820 Boston's drainage network was plagued with problems. Municipal incorporation provided Boston with the means and opportunity to rectify these shortcomings. The city sewer system rapidly expanded after 1823. Despite their origins in private storm drainage, by the 1830s these pipes were widely seen as the solution to the many problems associated with privy vaults.

The introduction of human wastes into the poorly designed drainage network ultimately increased sanitation problems, but even these problems paled in comparison to the old privy vaults. In many nineteenth-century communities in the United States and Europe, scavengers hauled wastes out of privy vaults and sold them for use as fertilizer on nearby farms. Frequently accused of doing their work sloppily, scavengers did spill trails of wastes behind them as they hauled their carts and barrels through city streets. Once public health officials began to associate human wastes and related odors with disease, these unpleasant aspects of privy vaults came to be seen as threats rather than mere inconveniences. Innovations in vault cleaning machinery, including sealed pumps and barrels, addressed some of these concerns. But these technologies developed in competition with the water-carriage sewer and never overcame the negative association with the old-fashioned, unsanitary scavenger. The final chapter on privy vaults closed with the introduction of running water. The resulting increase of waste water pouring into privy vaults hampered cleaning practices and increased the volume of waste water beyond the capacity of the disposal system. Thousands of gallons leaching from vaults saturated the soil, flooded cellars, contaminated wells, and threatened the public health.[9]

In Boston, as in so many other cities, the weaknesses of the privy vault system were most apparent in poorer neighborhoods. Ironically, while these wards provided the impetus for the adoption of new technologies, the poor had to wait many years before new waste disposal networks reached them. Boston's laws regulating the construction and maintenance of privy vaults gave landowners and their tenants equal responsibility for waste disposal.[10] Because these laws did not establish a clear responsibility for sanitation in rented houses, they actually delayed the construction and timely cleaning of waste facilities. The city did pass laws compelling privy vault construction and could order vaults cleaned at the owner's or tenant's expense. But municipal health officials and police enforced these measures unevenly. Thus many parts of the city, particularly poorer neighborhoods with lower rates of property ownership, continued to experience saturated soils. Health surveys that found the greatest incidence of disease in wards with sodden soil, dense settlements, low incomes, and high immigrant populations reinforced the association between overflowing privy vaults, poor soil drainage, and ill health.[11] These findings attested to the limitations of private sanitary initiative.

In response, the Boston city council passed a law in 1833 that permitted landowners to drain privy vaults into nearby sewers.[12] Roof drainage and surface waters from private property were also permitted in the sewers; the increased flow from these relatively clean sources was intended to flush the system. The city council hoped these changes would reduce the liquid content of the vaults, facilitating vault maintenance and improving general sanitation. The law transformed Boston's storm drainage system into a de facto sanitary sewer system extremely early in the city's history. Thus Boston's residents were permitted to use the sewers for human wastes several decades before technological developments made water-carriage sewers an efficient, accepted aspect of urban infrastructure.

For most of the nineteenth century, neither new regulations nor increased municipal authority over Boston's sewers had much effect on drainage in poor neighborhoods. The Boston sewer department, until the 1870s, built sewers primarily in response to petitions from private citizens who wished to improve drainage in their neighborhoods. City coffers paid only a small portion of the costs for this construction. Most of the funds came from "improvement assessments" added to the property taxes paid by those with lands abutting the new sewer line.[13] Although the wealthier members of society could take advantage of municipal sewer construction, poor residents could not afford sewer assessments. Meanwhile, renters had neither incentive nor means to participate in construction funded by property taxes. Indeed, Boston's poorest residents lacked privies entirely. According to Dr. C. E. Buckingham, a physician at the Boston Dispensary, in their households "excrement [was] thrown into the yard, and even under the lower floor of the houses."[14] Sewer construction was thus likely to offer such communities little benefit. Even in the absence of specific information on the financial status of the people who requested sewer construction, it is possible to surmise from tax and fee structures that wealthier Bostonians were more likely than their poorer neighbors to enjoy state-of-the-art sanitation. The city's approach to funding and planning sewers reflected demand by the middle and upper classes, rather than need among the poor.[15] Because poorer neighborhoods had higher disease rates, building sewers there could have made a tremendous difference in overall public health. As it was, city sewer construction failed to significantly decrease disease; this failure in turn, made the city look ineffective.

The shortcomings of both the old private sewer network and new sewer department policies emerged soon after the city took over sewer construction. In 1848 Buckingham noted that up to two thirds of Boston's sewers had so filled with mud that they no longer provided any drainage. Moreover, the city provided sewers only in the largest streets. Private landowners still had to build their own sewers in streets less than thirty feet wide.[16] This policy reinforced the inequity in the distribution of sewers throughout the city. The simple transfer of responsibility for sewers from private to public hands changed Boston's sanitation very little. Although improved construction standards established by the sewer department prevented some of the worst problems associated with faulty sewer pipes, these standards did little to extend drainage to older, poorer neighborhoods or to those with narrower streets. The city did not require sewer connections to all Boston buildings until 1871;[17] at that time nearly half of the city's human wastes ended up in privy vaults or cesspools. The continued use of privies and the faulty pipes installed before the city established its construction standards prevented municipal drainage from having any dramatic effects on public health.

The Cochituate Waterworks

For most of its history, Boston's ground and surface water had supplied the growing town with water for all its domestic and commercial needs. Wells and cisterns were extremely vulnerable to contamination from overflowing privies and old, leaking sewer lines. Because municipal sewer construction could not address all these problems and therefore failed to reduce disease significantly, Bostonians soon turned their attention to water itself. In time, their efforts yielded the Cochituate Waterworks, a municipally owned system that offered more direct benefits to Boston's poor than had public sewer construction. Proposals for the waterworks grew out of the same sanitation impulse that drove sewerage, including the vague hopes that services would reform poor communities. Acquisition and construction of Cochituate, however, marked an even greater departure from legal and institutional precedents than did the creation of Boston's sewer department. In this way, demand for improved water service and protection from epidemics precipitated dramatic growth in city government. Within a few decades, municipal powers far exceeded those established in Boston's original charter.

The history of Boston's municipal water supply began with several eighteenth-century efforts to improve fire protection. In Boston, as in most early cities, fire loomed as an ever present danger. Wooden buildings, narrow streets, the extensive use of open flames for heat and light, and the reliance on volunteer fire fighters combined to make seventeenth- and eighteenth-century cities especially vulnerable to destructive conflagrations. In 1676 one of Boston's first major fires destroyed forty-five buildings, the North Meeting House, and a number of warehouses. Flying embers from this fire also threatened to set nearby Charlestown alight. Three years later, 150 buildings and several ships burned. In 1760 the Great Fire reduced 174 dwellings and 175 commercial structures to ash, leaving 220 families homeless. Thirty years later, another fire burned 96 dwellings and caused $209,000 in damages.[18] In those early years, fire protection, like water supply and sewerage, was a product of private initiative. In 1652, the Water Works Company built Boston's first water system, a wooden reservoir and conduit to improve fire fighting in the central business district. These structures may have survived into the 1770s, but ultimately could not protect the city from one destructive burn after another.

It was not until almost one hundred-fifty years later that a second private water enterprise arose in Boston. The Aqueduct Company, founded in 1795, developed a more sophisticated water supply network complete with underground water lines and a reservoir in the suburbs. The Aqueduct Company ran nearly forty miles of pipe from Jamaica Pond in Roxbury to buildings in the southern and western sections of the city. This system also offered fire protection; the company complied with a legislative order to install twenty-nine hydrants along its water lines. Even so, the company's limited service area left most of Boston's commercial and residential districts without hydrants. To protect the remaining territory, the city constructed eight reservoirs or public wells, and secured the use of twenty-five additional private wells, reservoirs, or water tanks for use during fire fighting operations. As needed, the city added new reservoirs to serve various neighborhoods. However, these early measures failed to provide adequate fire protection. Fire and disease prevention provided all the motivation that Bostonians needed to improve their water systems in the early nineteenth century.

The first proposals for a municipal water supply in Boston emerged shortly after incorporation. Advocates of public water saw Boston's new municipal status as an opportunity to expand public services beyond those outlined in the charter. They cited reports of Boston's dire water shortages, the success of New York's Croton Aqueduct, and the disappointing results of sewerage construction in their arguments in favor of a city water network. In 1825, contamination of wells by privies and the need for improved fire fighting capabilities renewed interest in municipal water proposals. Entrepreneurs tried to take advantage of the growing market for better water by building small private systems, but these never caught on; few Bostonians were willing to pay for water service when they could make do with unpalatable water from their wells. Even after they admitted that they needed new supplies, high construction costs and fears of excessive government spending and power delayed the public waterworks for many years.

In 1826, Josiah Quincy, Boston's second mayor, made water a chief goal of his administration, declaring that local physicians had "urged this topic upon me . . . on the ground of health in addition to all the other obvious comforts and advantages" of a municipal water system. John C. Warren of the Board of City Physicians argued that the overall benefits of public water supply outweighed the expenses, that pure water would improve public health, and that the poor quality of existing supplies represented a threat to the continued development of the city. Warren supported his conclusions by citing cases in which physicians cured disease merely by replacing patients' contaminated water with pure supplies. In 1825 Quincy had appointed a committee to study Boston's water situation and the possibilities of creating a municipal water supply.[19] Although Mayor Quincy championed the water system, nothing came of the recommendations of this first Boston water committee.

In 1827, in response to continued calls for water improvements and continued city inaction, George and Thomas Odiorne proposed a private water supply for Boston. Working in the tradition of the Boston Conduit and the Boston Aqueduct Company, the Odiornes offered to build a water supply using their rights to Spot Pond, north of the city. The city water committee rejected the Odiornes' proposal within the year, but the topic remained on the city's agenda off and on throughout the 1830s and 1840s.[20]

An 1834 survey of Boston's wells provided ample evidence that the city needed improved water supplies. Laommi Baldwin, one of a family of prominent engineers, identified widespread problems with the city's groundwater. "The wells in town," he reported, "are polluted by the dirty water at the surface . . . mingling with the veins below." Wells that were not fouled from above were "adulterated by mixture with little streams of sea water" underground. In other places, salt and fresh water springs occurred so close together as to prevent drilling a well to tap the fresh water. Baldwin did find that most of the wells in the city (2,085 out of 2,767) provided drinkable water. But only seven of the wells delivered water soft enough to use for washing. Of all of Boston's wells, 427 failed regularly, 62 were either contaminated by household wastes and privy vaults or could be considered "nuisances." Baldwin required an additional 134 wells "brackish, bad, . . . or turbid."[21] In sum, Baldwin listed fully a quarter of Boston's wells as substandard. Bostonians' private, independent supplies were gradually but clearly failing.

Baldwin suggested that the city remedy these problems with a large public water project. He rejected both private water development and all of the small water supplies recommended by earlier reports or offered to the city by private investors. Baldwin preferred Lake Cochituate, a 600-acre lake located some sixteen miles west of Boston.[22] Cochituate appealed to Baldwin for a number of reasons. Larger than most of the other possible resources located within twenty miles of the city, the lake could provide ample water for Boston for many years to come. Surrounded by undeveloped land, Cochituate seemed protected from industrial and urban pollution. Cochituate also lay high enough above Boston to allow water to flow into the city under its own power. Gravity is still by far the most inexpensive and reliable way to move water. Ease of transporting water to the city, water purity, and reservoir size were all critical to Baldwin's recommendation.[23] Despite the dire conditions of Boston's wells described in Baldwin's report, the city council took no decisive action on the water question in the 1830s.

By 1836, Boston's water supply debate had evolved into a contest between low taxes, a small government, and private enterprise on the one hand, and increased government action on behalf of public health and economic growth on the other. A number of local leaders, Mayor Theodore Lyman among them, also cited the value of running water for fire

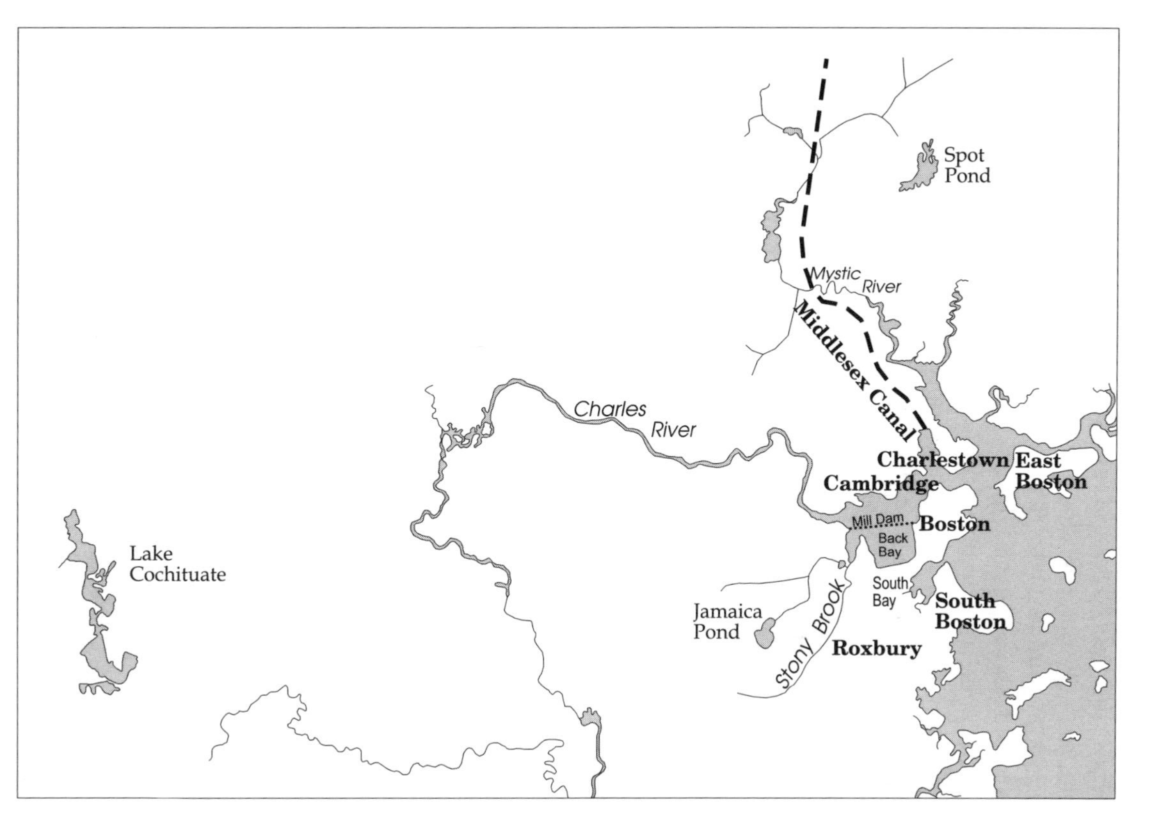

Municipal water in Boston. Boston's municipal water system tapped Lake Cochituate. Before that system was built, entrepreneurs offered to build private waterworks drawing on the Middlesex Canal or Spot Pond. Beginning in 1795, the private Aqueduct Company sold water from Jamaica Pond to a few Boston households.

protection.[24] During the 1830s and 1840s, private water proposals surfaced whenever support for water development peaked or opponents of public water replaced proponents in the mayor's office. George Odiorne would remain in the public eye, pushing his Spot Pond proposals as late as 1844. Similarly, in 1845, the proprietors of the Middlesex Canal presented their obsolete navigation channel as a water supply and conduit for Boston.[25] The persistence of these schemes reflected the lack of consensus among city leaders regarding how they should improve Boston's water supplies. It also suggests that the private water developers sought to take advantage of both frustrated support for a municipal water supply and the cautious fiscal policies of public water opponents.

Throughout these water discussions, Boston's mayors played critical roles in promoting or delaying the water project. Boston's municipal officials had to seek reelection every year; short terms and frequent elections interrupted complicated planning processes and diffused the momentum needed to gather public support for expensive projects. Nevertheless, whenever the mayor pushed water development, the city moved decisively towards approval. But if the mayor focused municipal attention on other matters, particularly fiscal retrenchment, water lagged. At several critical junctures, just when Boston seemed about to proceed with the waterworks, voters elected a mayor opposed to the project. Then, even petitions from the City Physicians about the urgency of introducing "wholesome water" could not spur the city council to take concerted action.[26] For example, the cautious financial policies of Mayor Theodore Lyman in 1834 and 1835 overshadowed concern for municipal water. Samuel T. Armstrong, in contrast, took a strong pro-water stance in 1836. Citing a cholera epidemic then striking New York, he advocated water surveys and directed the city council's water committee to examine whether public or private enterprise could best provide water for Boston.[27] But Armstrong took no further action.

The election of alternating pro- and antiwater mayors suggests that Boston voters remained unsure of the limits of public authority and hesitated when called upon to give the city officials broad discretion. However, during this period public votes specifically on the water issue revealed growing support for municipal action. In a public referendum in August 1836, ninety-four percent of Bostonians attending a special meeting approved municipal water.[28] That fall, voters elected a strong water advocate, Samuel A. Eliot, to the mayor's office. During three successive Eliot administrations in 1837, 1838, and 1839, the city made substantial progress towards municipal water. Eliot labeled water one of the city's most pressing issues in each of his three inaugural addresses, raising expectations that the water system would decrease vice and increase property values in the city.[29] In 1838 Eliot proclaimed himself firmly opposed to water proposals by private corporations; in 1839 he chided the city council for its inaction on the subject. Voters reflected Eliot's enthusiasm for a public water supply, voting three to two in favor of a publicly funded system.[30]

Eliot's advocacy for the waterworks may have backfired. Jonathan Chapman, the mayor elected in 1840, 1841, and 1842, took a strong anti-water position. The chief priority of his administration was to reduce municipal debt and expenditures. He called water supply a public expense that should be delayed or rejected because the public remained too divided on the issue.[31] By 1844, however, public sentiment appeared divided no longer. In public meetings, seventy four percent of voters approved both the idea of public water and the Cochituate project as outlined by Laommi Baldwin in 1834. Thomas A. Davis, mayor for part of 1845, repeated Chapman's cautious sentiments, but his successor, Josiah Quincy, Jr., proved as strong an advocate of the public water project as his father had been.[32]

Public health and sanitation benefits figured prominently in nearly all discussions of the water project. Significantly, even opponents of Cochituate rarely disagreed with the public health claims made by water advocates. One public water proponent, John H. Wilkins, wanted to take municipal water even further, calling on the city to "distribute the water *for domestic purposes*, free from charge."[33] Physicians focused on the dangers of contaminated water and argued that a new supply would dramatically reduce suffering and disease in the city.[34] High mortality and morbidity, they claimed, interfered with Boston's prosperity and commercial growth in a number of ways. Poor public health stained the city's reputation. In addition, because disease made the poor all the more wretched, poor sanitation increased the cost of public support for the indigent, reducing resources available for expenditures that might attract business. Furthermore, the specter of devastating epidemics terrified most urban residents of the middle and upper classes. Ironically, Boston's earlier public projects had not delivered services in ways that significantly improved conditions for those most vulnerable to infectious disease. Even so, the city council adopted public health as a major justification for increased municipal powers, responsibility, and expenditures.[35]

Many water advocates thought that the benefits of public water extended far beyond questions of health. Social reformers, for instance, believed that the water project would help them achieve their broader goals. For example, the Total Abstinence Society, a temperance organization, endorsed the public water plan as the best way to ensure the widespread installation and use of drinking fountains.[36] Ample water, they contended, would give thirsty Bostonians an alternative to liquor.

Of course, financial considerations held considerable sway over decisions on the water issue. Investors in the Boston Hydraulic Company and the Boston Aqueduct Company opposed a public system because of their financial stake in private water projects. Similarly, owners of those water-dependent industries likely to be adversely affected by any particular water proposal either opposed that specific project or public water generally. Mill owners along the Concord and Sudbury Rivers and in Framingham opposed the Cochituate project because it would decrease the water power available to their factories.[37] Similarly, the proprietors of the Middlesex Canal, downstream from Cochituate, objected to the municipal project on the grounds that the legislation authorizing the canal conveyed water rights to the canal that neither the state nor Boston should rescind.[38] In some cases, opposition to one project intertwined with investments in or benefits from another. For example, Caleb Eddy, who owned shares in the Middlesex Canal and served on the executive board of the Spot Pond Aqueduct Company in 1845, opposed Cochituate. If he and his colleagues had convinced Boston to accept the Spot Pond option, he would have gained twice—once from the returns from Spot Pond, and once because of the continued viability of the Middlesex Canal.

While public water proponents dismissed private water advocates as self-interested and profit-oriented, their opponents argued that public water represented an illegal expansion of government and an infringement on private property rights. Some argued that the city did not have the authority to build a water system. They feared that the water project might lead to further, unreasonable expansion of governmental power "to carry out some new project for supplying heat and gas from the interior of the earth; or to borrow money . . . to supply clothing for the poor."[39] Private water advocates Lemuel Shattuck and George Odiorne emphasized the injustice of using tax revenues for a project that, they claimed, effectively charged residents for water that they did not need or might not receive. William J. Hubbard argued at a legislative hearing that Boston was requesting too many powers and that the city's project would interfere with the rights of other towns and of mills and other industries.[40] Jonathan French, in his remonstrance against the water project, complained about the "invasion of private rights" inherent in a public water project.[41]

Some industries, particularly those that depended less on private ownership of river rights or expected to benefit from improved supplies within Boston, actively supported municipal water. These included a number of real estate development groups such as the South Cove and Mill Pond companies, which were engaged in filling and building on marshlands and tidal flats around the city.[42] As their positions demonstrated, economic self-interest played a substantial role in separating water supporters from opponents. In the end, however, neither the high cost of the water system nor the difficulties of distributing the costs of the project deterred a majority of Bostonians from voting in favor of public water at several public meetings and at the final water vote. Bostonians proved eager to embrace expanded government as long as it promised to provide services that they deemed necessary.

In 1845 after two decades of debate, one final shiver of hesitation stood between Boston and its municipal water works. That year voters narrowly rejected legislation passed in the General Court that was intended to empower Boston to proceed with Cochituate, despite a call for immediate action by Boston's second Mayor Josiah Quincy.[43] But when a second water act came before them in 1846, they approved it thirteen to one.[44] The 1846 proposal received greater support because it specified that Boston was to develop the Cochituate source for its water project. Previous legislation had left the choice of water source up to the discretion of the city council which, voters feared, would invest in a second rate project for the profit of private interests. Their reactions to the two water bills not only demonstrated that voters wanted to retain their influence over elected officials, but also their relatively new assumption that public investment should benefit the general public rather than the elites or special interests.[45] This distinguished Cochituate from earlier public enterprises built in response to petitions from and intended to benefit Boston's elite.

The 1846 Act for Supplying the City of Boston with Pure Water removed the final barriers to Boston's municipal Cochituate Waterworks.[46] Although mayoral support played a major role in the timing and success of the Cochituate proposal, many other interests had to fall in line as well. Opinions on the Boston water system developed over the two decades during which the water issue lay before the Boston public. Proponents insisted that the public health and development benefits of the system outweighed problems associated with increased public expenditure. They further insisted that water was vital to the social fabric of Boston society and lay firmly within the responsibilities of municipal government. Opponents on the other hand, focused on the burdens of debt and taxation and on the merits of private services. Of these issues, the questions of public health, municipal authority, and public versus private water development would prove most critical.

Approval of the Cochituate project firmly established both the primacy of urban resource needs over private economic interests and the authority of municipalities to provide their residents with basic necessities. Successful public works had similar effects in a number of other eastern cities. In each case, sanitation projects represented a crucial stage in the development of municipal government, expanding the territorial reach and types of powers granted to local officials. Boston's semiautonomous water board would prove to be one of the most important legacies of this expansion of municipal authority. Justified as a mechanism for expert administration of the waterworks, this new agency introduced political reform into public works construction. Because appointed officials and water experts rather than elected representatives staffed the Cochituate Water Board, the waterworks decreased voter supervision of policy. Nevertheless, the success and popularity of the waterworks established the voters' confidence in the water board in particular, and in semiautonomous agencies in general. As a result, the board would serve as a model for subsequent city and regional endeavors ranging from airports to parks.

For decades after Boston completed the Cochituate Waterworks, advocates of public enterprise cited the benefits of the water network to justify other public expenditures. Public reactions were also enthusiastic. The extensive water distribution system brought pure water to every section of the city and, incidentally, proved a powerful bargaining tool when Boston sought to convince neighboring towns to accept annexation. All too soon, however, Boston confronted the inadequacies of municipal construction. Services remained unevenly distributed throughout the city. Waste disposal, in particular, suffered. The old pipes proved ill-suited to the new regime of water-carriage sewage; they concentrated wastes along the waterfront and in stagnant, foul eddies on all sides of the city.

The water supply system, because it was built from a single plan, escaped some of these problems. Nevertheless, the Cochituate Waterworks soon ran into difficulties of its own. As Bostonians installed appliances to take advantage of running water, water use soared beyond projections. A few years after Cochituate water reached Boston, the city found itself facing unexpected water shortages. By the 1870s, these problems forced the city to expand its services. In the process of seeking better sewage disposal and additional water supplies, however, Boston had to compete with its suburbs for water resources. Although the city did significantly expand its water and sewer systems in the 1870s, this second round of municipal enterprise generated even more contention than the first because the city found itself constrained by its own territorial boundaries.

East Bay: The Private Service Tradition 1850–1905

When it came to public utility development, Boston had a distinct advantage over many western cities. Massachusetts' tradition of public water rights and the causal relationship in Boston between public works and public health paved the way for the expansion of government authority over public services. In contrast, western development took place in an atmosphere of private resource distribution. As a result, California cities had far less success than their eastern predecessors as they struggled for authority to regulate and operate public services. Private entrepreneurs profited handsomely from water supply, transportation, and power networks. They resisted efforts to limit their operations and their income. Their very imperviousness to public authority inspired repeated attempts to assert municipal and state control. In the East Bay, regulation of water and other utility companies remained extremely limited until the early twentieth century. Nevertheless, East Bay campaigns to rein in private water companies increased the power of local government and established important precedents for subsequent water battles, much as Boston's earlier and more successful water and sewerage projects had.

Private enterprise did not dominate all service responsibilities. Indeed, residents expected local officials to provide sanitation, to lay out streets, and to fight fires. These services offered little opportunity for immediate profit and had long since been ceded to public authority throughout the nation. Moreover, sewer lines and paved streets frequently could wait until residents agreed that the city government should undertake improvement projects. In contrast, scarce water supplies made individual wells or cisterns unreliable for most urban residents and forced many communities in arid locales to seek centralized water development long before they were ready to tax themselves for these services. In the absence of a consensus in favor of public water development—and the increased municipal spending that came with it—private water companies took charge of supplying water to thirsty urban residents. Because residents had few alternatives to purchasing water, private water companies proved attractive and profitable ventures. These companies used public-spirited rhetoric, political influence, and substantial financial resources to defeat proposals for government regulation or takeover. A comparison between the private utilities' monopoly of waterworks and the city government's responsibility for sewerage illuminates the limits of public authority in California.

Antimonopoly sentiment lay at the heart of nearly all efforts to exert government control of private utility companies in California. This sentiment reflected much more than public resentment of corporate profiteering. Early criticism of railroads emerged in response to slow economic growth and resentment of state and federal subsidies of wealthy railroad magnates. Agricultural communities found the railroads' rate structures unfair and inefficient; shipping prices seemed to favor the wealthy over modest farms and companies. Southern Pacific's influence over local and state politics and the failure of early regulations did little to mollify railroad opponents. Nativists blamed railroad companies' practice of hiring Chinese laborers for the huge influx of Asian immigrants to California; thus the violent anti-Chinese biases of the era added to anti-utility sentiment in nineteenth-century California.[47] Furthermore, as Donald Pisani has noted, some individuals with huge land and water holdings blamed the railroad for local economic travails in an effort to deflect attention away from their own roles in the difficulties of homesteaders. The railroads' reluctance to sell its huge land holdings to tenants and squatters further alienated many Californians.[48] Agitation for railroad regulation began in the 1870s in response to these varied complaints, but inadequate legal authority and precedent, combined with unfavorable judicial decisions, hindered these efforts until the early twentieth century.[49] Although most antimonopoly rhetoric targeted the railroads, urban residents directed similar criticism at other private utilities as well. This was particularly the case as economic growth flagged in the late nineteenth and early twentieth centuries, and as utilities seemed to emulate the railroads' pricing and political practices.

Oakland's early efforts to establish municipal control of public works was typical of California's struggles to rein in private utilities. The city was incorporated in 1852 as part of a land speculation scheme; many of the city's services developed with the explicit goal of increasing business opportunities. As the northern California terminus of the Southern Pacific Railroad, a company that came to control the city's entire waterfront, streetcar system, and ferry connections to San Francisco, as well as all rail shipping in and out of the city, Oakland was the epitome of a railroad town. Frustration with the railroad and other monopolies ran high.

Like the railroads, Oakland's early water companies used their monopoly to their best advantage. They gained control of available water sources early in Oakland's history. They used their control not only to guarantee their revenue but also to stave off public challenges to their monopoly. Although water users frequently complained about the poor quality of the water they received, high costs and poor service caused an even greater stir. And while the city's charter clearly granted Oakland the authority to build streets and sewers, water, because it was a salable commodity, received no such consideration. Oakland's combination of public and private services contrasts sharply with Boston's authority over both water supply and sewage disposal. Boston successfully established control over water resources because water in Massachusetts was at once less valuable and more closely associated with public health.

Oakland Sanitation

Services and utilities in Oakland fell into two broad categories: revenue-generating and non-revenue-generating activities. Water, transportation, power, and communications made up the first category and were privately developed. The companies that built these networks would present California cities with their greatest challenges to public authority. Activities that did not generate revenue, such as street and sewer construction, law and building code enforcement, fire fighting, street cleaning, and rubbish disposal, did not attract private investment. They were, however, no less necessary to ensure profitable industry and community growth. Like many other city charters written in the 1850s, Oakland's original charter granted the city council and mayor authority over non-revenue-generating functions.

Of course, the limits of East Bay municipal authority were not determined solely by the marketplace. Oakland residents brought a number of expectations to the task of creating a municipality; some of the powers outlined in Oakland's first charter, including responsibility for sewerage, reflected the experience of other communities rather than conditions in the East Bay. As a result, sewerage proposals came before the Oakland public accompanied by less crisis rhetoric or political controversy than they did in Boston. Few in Oakland feared that their waste disposal practices would cause disease because the East Bay's dry climate and good drainage, together with San Francisco Bay's strong currents, rendered shoreline sewage disposal fairly innocuous. Moreover, because Oakland functioned more as a residential suburb of San Francisco than as an independent metropolitan core, the East Bay lacked much of the ethnic and class divisiveness seen both across the Bay and in Boston. So covert accusations of moral and physical corruption among the poor, which colored the debate in Boston, had less relevance in East Bay politics. In cities like Boston, epidemics and political interpretations of urban pollution helped establish early-nineteenth-century public authority over sewerage; in Oakland, the inherently unprofitable nature of sewerage and boosters' faith that improved services would secure their city's fortunes prompted municipal officials to take responsibility for sanitation.

Oakland's early history reflects the speculation and appropriation of resources that so tainted California's first years as a state. In 1820, Don Luis María Peralta received most of what is now Alameda, Contra Costa, and Santa Clara counties in recognition of his military service to Mexico. Peralta's hold on the land did not last long, however. In the 1840s squatters and speculators began to encroach on the Peralta rancho. Beginning in 1846, Peralta and his sons rejected a series of proposals for urban development in the East Bay. But the Peraltas' refusal failed to reverse the flow of squatters. In 1852, Horace Carpentier founded the city of Oakland on land he did not own. Not until 1854 did Carpentier and his fellow speculators purchase land from the Peraltas to make their settlements legal.[50]

In the early years, Oakland, like other western cities, saw railroads as central to local prosperity. So, in 1867, when Southern Pacific began looking for a northern California terminus, Mayor Samuel Merritt joined forces with Carpentier to attract the railroad to Oakland. At the time, Carpentier was a director of Central Pacific Railroad, one of the subsidiaries of the Southern Pacific. Together, they convinced the Oakland city council to abandon all legal claims to the waterfront. Next, they founded the Oakland Waterfront Company and promptly sold five hundred acres of prime land to Western Pacific Railroad, a subsidiary of Central Pacific.[51] The railroad quickly developed piers and other facilities to carry goods from their trains by freight and ferry to San Francisco. Oakland became a major transshipment center for goods bound to and from San Francisco, the gold fields, and points south and east. The emphasis on growth demonstrated by California's dedication to squatters' rights, at the expense of Mexican, Chinese, and Native American residents of the state, and the eagerness with which communities embraced machinations like those undertaken by Carpentier and Merritt, would dominate public policy for many years.

In 1869, Oakland undertook its first formal sewerage construction; the project was intended to attract business and help Oakland compete with its neighbors. Up to that time, residents and businesses disposed of their wastes in cesspools or local waterways. Although these practices were identified with epidemics and urban filth in Boston, in Oakland they hardly caused a stir. The engineers designing the project called for the construction of two main sewers, each about three miles long, to collect wastes along each side of the city. Although they believed that Oakland could not yet afford settling tanks or an outfall long enough to remove wastes from Oakland's shores, the engineers designed the five-foot-high brick sewer mains with such later additions in mind. They observed that "no material injury . . . to the general health of the City will result from discharging the present sewage directly into the creek," but diverting wastes from the creek remained a long-term goal.[52] As they noted: "Many of the older cities have recently abandoned the time-honored custom of discharging their sewage matter directly into the adjacent water courses, having found that the effluvia arising from their polluted waters was not only highly offensive but that the resulting miasma increased, in a very marked degree, the ratio of mortality."[53]

Significantly, Oakland's engineers referred to public health problems in only the most general terms, and then only in connection with the experience of other cities. In Oakland, public health did not, in 1869, justify expensive construction; sewage disposal practices were not seen as constituting a "material injury" to the "general health of the city." Instead, Oakland began to build sewers to emulate and to compete with larger cities.

The 1869 sewers marked only the beginning of Oakland's sanitary endeavors. Oakland undertook additional sewer construction in 1876 and 1885. In 1887, Mayor W. R. Davis noted that the original sewers needed repairs almost constantly. He suggested that the city rebuild the whole sewer network "once and for all." Davis, however, did not stop with existing needs. In his annual message, he urged the city council to approve further sanitation construction to provide for future growth.[54] Mayor Anson Barstow repeated this sentiment more explicitly in 1902, when he advocated comprehensive urban improvements, including redesigned thoroughfares and new parks as well as sewerage. Barstow promoted such public expenditures as means to enhance Oakland's "prestige" and "progress."[55] Clearly, these projects reflected the concerns of boosters and developers more than they did those of health officials or social reformers.

The history of Oakland's early sewer development reveals several important features of public works expansion. Where Boston tended to build sanitary networks in direct reaction to approaching epidemics or existing sewage nuisances, Oakland initially anticipated sewage demand. The smaller city used public works to emulate and to compete with larger communities; so these projects appealed to Oakland's boosters. Oakland's aggressive sanitation program is striking, given the city's limited authority over water resources, and given the long delays before Oakland residents supported public development of water services.

Oakland's Private Water Monopolies

Oakland city officials had little influence over water service because individuals and private companies claimed East Bay water resources long before the city was ready to build a public waterworks. Boston faced no such barriers because Massachusetts law reserved Great Ponds for public use. Simple abundance did contribute to generous provisions for public water rights in Massachusetts. Pre-nineteenth-century legal traditions intended to protect common uses of streams, however, did even more to protect public interests. By the mid-nineteenth century, the demands of industrial development placed enormous strains on public or common water rights even in Massachusetts. California was settled in the midst of this new emphasis on private rights; in Oakland, private development and ownership of natural resources held sway.

California's early history was marked by rampant speculation and aggressive appropriation of natural resources. Even before the gold rush, land, otter pelts, leather, and beef tallow attracted traders and migrants to the West Coast. Of course, gold caused the greatest conflicts over resource distribution after 1849. In the gold fields, miners improvised land and water regulations, laying the foundation of California's water law.[56] Elsewhere the new arrivals appropriated Californios' ranchos, despite pledges to uphold Mexican land grants. Soon, squatters' camps became recognized, even incorporated, cities. After California became a state in 1850, costly litigation over land ownership further disenfranchised many Californio landholders. In striking contrast to Massachusetts, questions of property and resource access in California revolved primarily around the distribution of private rights. A history of private appropriation of and competition for natural resources colored urban as well as rural development in the state.

In keeping with California's resource policies and the distribution of resources into private hands that characterized western settlement, Oakland's original charter did not provide for public water development; for many years creeks and shallow wells provided sufficient water for most local needs. Soon, however, the growing community found itself in need of more elaborate water systems. As in so many other parts of California, a preference for limited municipal powers and low taxes prompted Oaklanders to turned to private water companies when they outgrew household wells. In 1858, the California legislature imposed minimum standards on private utilities; regulations governed the way corporations could claim water and land and clarified the legal relationships between private utilities and municipal governments. The legislation required water companies to furnish free water for fire protection and to provide water to all local residents who applied for service. A commission appointed jointly by the water companies and city officials set water rates.[57] Significantly, authority over California water utilities resided in state rather than local hands. The jointly appointed water commissions allowed municipal government only token oversight over water rates and policies. To many opponents of the utilities, the commissions appeared to institutionalize utility influence over municipal government.

The 1858 legislation allowed investors to buy up water rights and hold them for later development or resale. The East Bay's first water companies were just such speculative ventures. In 1860, for example, the Alvarado Artesian Well Company incorporated and claimed rights to a large aquifer south of Oakland.[58] This aquifer eventually supplied substantial amounts of water to Oakland, Berkeley, Richmond, and Alameda.

But neither the Alvarado Artesian Well Company nor two other companies that were granted franchises during the next six years ever delivered water to customers. Instead, investors recovered their money when they later sold their water rights to active concerns.

The Contra Costa Water Company, founded in 1866 by Anthony Chabot, was one of the companies that would eventually buy up water rights throughout the East Bay.[59] Chabot was not new to water development. He had helped develop hydraulic mining in the Mother Lode, and in 1858 he built San Francisco's first water supply. In Oakland, Chabot began laying pipe almost as soon as he secured his franchise from the Oakland city council. The following year, Chabot first pumped water from a private well into these pipes. By June 1867, the Contra Costa Water Company was in full operation, delivering water from a reservoir on Temescal Creek.[60] Over the next decades, Chabot extended both his water system and his monopoly of local water sources.

Oakland grew quickly, and residents soon found Chabot's Contra Costa Water Company lacking. By 1871, after only five years in operation, the company was receiving frequent complaints about water quality. When Chabot sunk wells near slaughterhouses south of the city, the uproar increased. In 1873, fed up with hard, murky water, Oakland residents began their first campaign for a public system. The following year, a citizens' committee recommended that Oakland build a large public water works. The committee praised municipal ownership as more economical than private water development, and called ample water crucial to Oakland's future growth and "self-protection."[61] This first statement in favor of public water foreshadowed many similar statements to come. For the next fifty years, East Bay advocates of public water would portray public waterworks as crucial for economic growth and condemn private water companies using the same rhetoric that Californians usually reserved for vilifying the railroad.

The 1874 report fell on receptive ears, both at City Hall and in Sacramento. The city council took up the issue but, instead of embracing public water, decided to help the private companies. In a move that resembled government subsidies for utilities elsewhere, the city council requested legislation to allow the city to aid the Contra Costa Water Company with waterworks construction. Chabot responded by offering to sell his company in exchange for municipal assistance with several pending water projects. As Chabot never quoted a price and so prevented Oakland from seriously considering the purchase, it seems likely that his offer was not sincere. In 1876, Chabot completed a new reservoir on San Leandro Creek, easing the water shortage and quelling public complaints.[62] Over the next decades, economic anxiety and temporary water shortages or declines in water service would prompt renewed calls for reform. Time after time, though, demands for public water would fade as soon as the crisis passed.

Periodically, other water entrepreneurs challenged Chabot by proposing new water supplies. Because these upstarts rarely had assets beyond water rights, however, their proposals revealed a desire to profit from water claims rather than to threaten the Contra Costa Water monopoly.[63] In the meantime, relations between the Contra Costa Water Company and Oakland remained essentially unperturbed.

The first serious challenge to Chabot's empire came some fifteen years after he completed the San Leandro Creek dam. In 1890, reformer George Pardee, later Oakland mayor and California governor, called upon the Contra Costa Water Company to filter water supplies provided to Oakland consumers. Alleging that poor water quality caused infectious disease in Oakland, he demanded that the city council either exert greater control over the water company or build a municipal system to provide pure water and to lower costs. To stir up opposition to private water companies, he reported that they charged large water users far less than they did smaller businesses. In fact, the average laundry or livery stable could expect the same annual bill as the California Cotton Mills or Southern Pacific's Coast Railroad depot. Pardee, who was to lead campaigns for public water development for the next thirty years, faced powerful opposition at this time. The city council, under pressure from the water company's president, Henry Pierce, and its attorney, J. C. Martin, defeated Pardee's proposal for municipal water, six to five.[64]

Just three years later, Contra Costa Water again had to defend itself, this time against a private competitor. In 1891, William Dingee purchased rights to the Alvarado aquifer. Two years later, he created the Oakland Water Company. These acts made him at once a nemesis to the Contra Costa Water Company and a champion in the eyes of the city council.[65] The council demonstrated its support with a franchise to provide water to fire hydrants located west of Broadway. The franchise gave Oakland Water the right to lay pipes through city streets and aided the company's efforts to develop a water network.

The hydrant franchise set off an era of cutthroat competition. The city council used the water companies' rivalry to its best advantage, offering hydrant franchises first to one and then the other company to induce them to lower the rates they charged the city. Some Oakland residents also switched water companies to reduce or avoid their water bills.[66] By 1895 the companies engaged in heated campaigns to discredit each other in the public eye. Dingee asserted that his groundwater supplies were naturally cleaner and more wholesome than Contra Costa Water's surface reservoirs because "so much filth and animal refuse is washed and carried into the lake by winter freshets."[67] His scientists found hog cholera bacteria in Contra Costa Water's reservoirs; they also reported that "the superiority of the Oakland Water Company's water as supplied to its *consumers . . . is proven beyond a doubt* by the total absence of injurious organisms in their water."[68] In 1897, the Board of Health corroborated some of Oakland Water Company's assertions and ordered Contra Costa Water to disconnect its pipes from the city's public schools.[69]

When scientific studies failed to sway public opinion decisively to one company or the other, saboteurs set to work. Mysterious holes punched in the flumes that crossed East Bay marshes let brackish water into Contra Costa Water's system. Elsewhere, someone tried to discredit the company by connecting a city sewer to a Contra Costa Water Company main. Contra Costa Water retaliated, diverting water into the bay from the Alvarado wells to reduce Oakland Water's supplies. Competition took its toll on water users as well as suppliers. As the companies tried to meet demands for their services and to keep up with shifting hydrant franchises, they built parallel, duplicate water lines through the city streets. To keep their costs down, they built the smallest possible water mains—pipes that would never be able to deliver the water pressure that the growing city needed. Falling water pressure angered residential customers and dangerously interfered with fire fighting.[70] The water company competition that had seemed to promise so much, had, by 1897, merely added to the city's problems.

In April 1897, Mayor Thomas proposed that Oakland end the chaos of water competition once and for all by developing a public waterworks.[71] In 1898, even as the city began investigating public development of wells in nearby San Lorenzo, the Oakland and Contra Costa Water companies negotiated a merger. Dingee took charge of a reorganized Contra Costa Water Company. He quickly raised its rates to recoup the losses of six years of competition and city-sponsored rate reductions. Next, Dingee negotiated the takeover of the two companies that had served Alameda and Berkeley.[72] This marked the beginning of a series of buyouts and consolidations through which Dingee secured nearly all the water supplies and water delivery systems in the East Bay. Contra Costa Water pursued its monopoly so successfully that East Bay cities soon had no other source of water, even if voters had approved a public system. At this point, California's tradition of private resource distribution utterly foiled the drive for public services.

Evidence of the water company's growing monopoly set off a spate of public water proposals. Citizens' committees, convened by local chambers of commerce as well as municipal officials, commissioned new water studies annually from 1900 to 1905, and sporadically until 1918. Meanwhile, prominent Oakland officials and business leaders argued that public investment could improve water supplies and service and free the city of the monopoly's control. Nearly all public water advocates decried Oakland's private water system as far inferior to model public networks in other major American cities.[73] During this time, public water proposals became fully integrated into local political discourse, much as the Cochituate water proposals had dominated mayoral campaigns in Boston from 1830 to 1846.

Diatribes against the water company were as common in East Bay politics as antirailroad rhetoric was in state-wide elections and jeremiads about political machines were in Boston. In 1902, for example, Mayor Anson Barstow claimed that political strife associated with the private water companies threatened "to destroy our municipal affairs."[74] The following year, Mayor Warren Olney asked "whether our city shall be ruled by her people, for her people, or by the Contra Costa Water Company for the benefit of the Contra Costa Water Company."[75] This rhetorical question reverberated with widespread anxiety that powerful interests threatened to overwhelm local governance. That same year, members of a water committee initially convened by the Barstow administration interpreted the disbanding of their group as proof that the water company did, in fact, control Oakland municipal government.

The water companies were accused of undue influence on Oakland's government on many occasions. The city council had some regulatory authority over the utilities to which it granted franchises. In the course of renewing the water franchises, for example, the council negotiated a thirty percent rate reduction in 1893 and a twenty percent decrease in 1895. In 1898 Oakland councilors argued that the newly reorganized Contra Costa Water Company should reduce rates again because water quality had declined. This time the dispute landed the city and the water company in court. The water company won and raised rates in 1901.[76] Significantly, voters dissatisfied with water rates or operations did not attempt to increase city authority over the water company. Rather, they saw the city council's failure to protect public interests as evidence that private corporations manipulated the city institutions. Public outrage over what was seen as illegitimate corporate influence in municipal politics eventually outweighed fears that a public water system would create new opportunities for graft.[77] Fear of the corrupting influence of private utilities increased support for government enterprise and regulation among groups that might otherwise have opposed the expansion of municipal authority. Although the precise mechanisms differed, reformers in both Massachusetts and California saw the expansion of public services as an ideal opportunity to transform local governance.

Of course, public water campaigns rested on more than the antimonopoly reform agenda. Whether discussing sewer improvements in the 1860s or public waterworks in the 1900s, Oakland boosters saw public amenities as vital to Oakland's continued growth. In 1902, for example, Mayor Barstow lamented that water rates disputes "retard[ed] growth and prosperity in our city."[78] For him, this was reason enough to promote a public water supply. Others agreed with his conclusions, arguing that only public ownership would permit the "far-sighted, efficient, economical and unhampered public service"[79] necessary for continued growth. Although similar enthusiasm for public enterprise existed in Boston, economic development more completely dominated public works debates in Oakland, largely because public health considerations had so little influence in the East Bay.

By the early 1900s, many business and political leaders in California and elsewhere rallied around the cause of economic growth and political reform. At all levels, leaders endorsed policies that were intended to foster growth by encouraging competition and regulating monopolies. Theodore Roosevelt's trustbusting, Hiram Johnson's antirailroad invective, and mayoral calls for a public water supply in Oakland were all intended to foster economic growth. Of course, as happened with other popular causes, the pro-growth rhetoric could be—and was—used by proponents of both private and public enterprise. To overlook either the political reform aspects or the anticipated economic consequences of Oakland's public water proposals (or Johnson's or Roosevelt's economic policies for that matter) would oversimplify the general appeal of increased public enterprise to most sectors of the population.

The joint endeavor to foster economic growth and protect government from corruption during this period led to a dramatic expansion of government authority in California. In Oakland, the transformation began with the city's efforts to mitigate the weaknesses of a private water system. When limited municipal regulation failed to live up to expectations, Oakland leaders sought other means to accomplish their goals, including litigation and public water development. Although public water proposals failed between 1873 and 1905, these early efforts established the economic priorities and the rhetoric that would characterize Oakland's water disputes for decades. In subsequent years, Oakland would try repeatedly to secure authority over water supplies. These efforts became considerably more complicated after the 1906 merger of the Contra Costa and Richmond Water companies. The merger created People's Water Company, a network of reservoirs and pipes that crossed county as well as city lines. Oakland had enough trouble regulating a utility within its own boundaries. Such regulation became almost impossible when the company's system extended beyond city limits. From this point on, public water advocates in the East Bay focused on regional, rather than municipal, proposals.

Municipal Authority over Water and Sewerage

By the end of the nineteenth century, both Boston and Oakland had developed a number of public services which demonstrated the particular strengths of municipal government. City leaders met little opposition as they assumed control over services that did not attract individual enterprise, those that cost more than they yielded in revenue or those that served clear, broad public goals. Sewers met these criteria, even though the two cities' goals for sanitation differed. Municipal projects that required public appropriation of water or other valuable, privately held resources met with resistance in both communities. Ultimately, the success of the water campaigns depended on local laws that favored public over private rights. But, as Oakland's experience demonstrates, voters did not leap to support expensive public projects if they saw viable alternatives to public development, or if they could not identify ways in which public enterprise promoted community goals better than a private system could.

Boston's first charter gave the city responsibility for local sewers because sanitation met all the criteria for public administration. When

Mayor Quincy issued Boston's first proposal for a public water system, however, even Boston could not claim all the conditions necessary for easy approval of such a project. Massachusetts laws did ensure public access to water supplies, but many residents still saw their wells and cisterns as adequate. Others did not fully accept the argument that running water would improve public health, did not see disease as a critical concern, or did not believe that municipal sanitation would protect them from epidemics. When municipal sewerage failed to reduce disease significantly, advocates of public water seized on the health issue for their campaign. Widely publicized studies that proved that Boston's wells were grossly polluted gradually undermined residents' faith in individual water supplies. Convincing the public that they needed both new water and public development took twenty years, but by the time voters approved the Cochituate Waterworks, Bostonians clearly saw the project as the best means both to implement public priorities and to respond to a local water crisis.

Oakland residents would not reach this sort of consensus for many years. Not only did private water companies provide a viable alternative to public waterworks, but East Bay residents could not agree that public development would serve public interests better than the private utilities did. Furthermore, Oakland officials found that their efforts to control the water companies were hampered by a legal system that favored private water rights. This legal tradition, consistent with federal policies during westward expansion, prompted mixed reactions among California voters. On the one hand, Californians generally approved of the way private enterprise supplied extensive services without high taxes. Many Californians celebrated private utilities as extremely efficient, and argued that state and local governments should administer public services in a more businesslike fashion. On the other hand, private ownership of resources and the emphasis on private enterprise created powerful corporate monopolies. In Oakland, for example, residents consistently blamed predatory monopolies not only for the defeat of public water proposals but also for a host of other economic and political frustrations. Oakland's love-hate relationship with both private industry and government enterprise significantly delayed the expansion of municipal powers.

Within the confines of their legal authority and the limits of public demand for new government projects, nineteenth-century cities significantly extended their services and responsibilities. Their early networks formed the physical backbone of later systems, while the institutional and political legacies of these projects shaped subsequent public enterprise far beyond municipal boundaries. Boston's Cochituate Water Board, for example, would serve as an institutional model not only for public works agencies in that city, but in Oakland and other cities across the country. Although Oakland's efforts yielded no institutions of such lasting significance, the East Bay's early water battles defined the goals and rhetoric of local public utility conflicts. In both cities, the first generation projects revealed the problems that would become chronic in later battles over public works expansion. Boston would continue to face competition from other communities. Time after time, Oakland would confront the power of private corporations and scarce natural resources.

As Boston and Oakland began to outgrow their first generation systems, they found that their own territorial boundaries proved the greatest barrier to improved public service. The second generation of services, although not yet regional in nature, would require cities to extend their power and infrastructure beyond municipal limits. The conflicts that arose from these efforts, together with the limitations of first generation municipal projects, eventually led to the widespread adoption of regionalism.

2 | Beyond Municipal Boundaries

Within a few decades of incorporation, Boston and Oakland had each established municipal authority over a wide range of services. By 1850 Bostonians had come not only to accept city services, but to expect them. In Oakland, public enterprise received less enthusiastic endorsement. Even so, by 1900 East Bay residents had come to see city government as the best source for numerous services. By regulating some businesses, subsidizing others, removing wastes, and facilitating water delivery, city officials made urban living easier and more attractive. Voters, in turn, rewarded their representatives for services delivered, and pushed them to expand public activities.

Ironically, however, the better services a municipality provided, the greater the demand for public works they inspired. In some cases, the lower prices, abundance, or convenience of city services prompted greater use of those services than anticipated. Ready access to running water, for example, nearly always led to a surge in water use that far exceeded projections. In some cases, resolving one crisis made others more obvious. Thus cleaning one filthy street might draw attention to the squalor around the block. Finally, population growth overextended services, so that even the best planned projects had a limited life span. Cities outgrew their services even faster if local laws, statutory ceilings on municipal debt, or opposition to public spending forced designers to build inadequate networks.

In many instances cities could meet public expectations by tinkering with existing networks or institutions. But when communities faced environmental crises, such small additions and adjustments did not suffice. Conserving water or installing new pipes could not fill an overdrawn reservoir. Likewise, building better sewer lines was of little avail if the sewage problem originated in another city or county. When facing service needs like these, hard-pressed cities adopted a number of strategies that involved looking beyond their borders. At times, cities annexed neighboring towns in order to coordinate administration of services or gain access to new resources. Interlocal projects, built cooperatively by two or more communities, promised to eliminate service duplications, pool funds, or help towns overcome the constraints imposed by municipal boundaries. In some cases, river towns even sued upstream communities to reduce river contamination. Unfortunately, each of these strategies reinforced the competition for resources and the conflicts among neighboring communities that ultimately impeded municipal efforts to satisfy public service demands.

Of all the efforts to overcome the constraints on municipal services, interlocal cooperation had the most promise. In general, interlocal cooperation took one of four forms: courtesy, contract, external, or symbiotic. These forms differed in distribution of benefits and degree of administrative cooperation. An example of a "courtesy" project is one city constructing a small section of water or sewer main in another's territory. Best suited for small pieces of a larger project, courtesy agreements abound in the Boston area because of the region's labyrinthine municipal boundaries and hilly territory. In the case of a "contract," one city buys a service from another. More widely used in California than Massachusetts, the contract is designed to save construction costs or gain access to resources controlled by another town or county. Cities build "external" projects outside municipal boundaries, but at their own expense and to meet their own needs. Because such projects offer no benefits to the communities closest to construction, external projects tend to confirm suburban fears of predatory and expansionist central cities. Finally, the "symbiotic" form of interlocal cooperation requires two or more cities to plan, build, and manage a single project together. This last form held the greatest promise for solutions to the complex sanitation and water supply problems in both greater Boston and the San Francisco Bay Area, but proved the most difficult to realize.

Interlocal agreements facilitated many of the public improvements constructed in the Boston area before 1890. Joint projects that included Boston itself, however, met with less success than those arranged among its neighbors. Boston's representatives rarely hesitated to use their city's political influence to secure water resources. While Boston's annexation campaign in the 1860s and 1870s had reflected suburban service needs as well as Boston's ambitions, its interlocal proposals were interpreted as an aggressive pursuit of municipal self-interest.[1] Many communities came to distrust Boston's motives whenever the city proposed a joint project. Although it did not diminish Boston's ability to construct courtesy projects, this distrust increased opposition to external projects and ruled out symbiotic proposals of any significance. By the 1880s, suburban reactions to interlocal proposals seriously compromised Boston's ability to meet citizen demands.Nevertheless, Boston area communities ultimately solved their environmental crises by extending public authority across municipal boundaries.

In California, Oakland and its East Bay neighbors eventually embraced the same approach as their Massachusetts counterparts, but they arrived at interlocal cooperation by a very different route. First of all, Oakland was never seen as the predatory monster that Boston was. In the Bay Area, San Francisco represented the greatest threat to local rule. As a result, East Bay towns pursued their own projects as a defense against San Francisco's expansionism. The existence of private water companies saved the East Bay from having to negotiate interlocal water projects but ultimately made the extension of public authority extremely problematic. Other California cities, Los Angeles in particular, had more experience with interlocal arrangements.[2] The contrasts between Boston's and Oakland's interlocal relations reflect the differences between a core and a peripheral city as well as the extent to which each city succeeded in exerting control over its services.

Although Boston and Oakland had very different records when it came to asserting public authority, ultimately proposals for interlocal cooperation served city residents no better than had single-city development. Rather than providing an easy way to address new service demands, these ventures increased suspicion in neighboring towns, created a race for scarce resources, and ultimately reduced the options for improving services. More important to the average citizen, even the most ambitious municipal schemes rarely resulted in long-term service improvements. Such disappointing results, from both internal and interlocal projects, shook the faith of many urbanites in their elected municipal officials. Failing services and unsuccessful improvement projects opened urban administrations to accusations of ineptitude and corruption. Eventually they led to the transfer of public services from elected to appointed officials, from politicians to engineers, and from municipal to regional or metropolitan administration. To understand that final transition, it is crucial to explore the ways in which Boston and Oakland attempted to address their water and sewer crises through the extension of municipal authority beyond city limits.

Boston: Expansion and Interlocal Competition, 1850–1880

By the 1850s, Boston city officials had firmly established municipal authority over a wide variety of public works. The Cochituate Water Board had completed the city's reservoir and water system. Acting—albeit slowly—in response to private petitions and health officials' recommendations, the city council directed other public agencies to improve sewers, water mains, streets, and other services. Nevertheless, Boston's projects could never keep up with demands for services nor solve the many local environmental problems that came in their wake. In the 1870s, because of this and a variety of political, social, and economic pressures, Boston leaders abandoned petition-driven construction for grander, centrally planned projects.

Three of the most substantial public works projects of the 1870s were the Sudbury River reservoirs, the Main Drain, and the Mystic Valley Sewer. Boston's water boards undertook the Sudbury and Mystic projects to increase and protect water supplies. A special sewer commission designed the Main Drain to facilitate waste disposal and divert sewage away from the waterfront. Meanwhile, the city sewer department built miles of new sewer lines to neighborhoods still dependent upon overflowing privies. These projects required a considerable outlay of funds. A national financial crisis and a twenty percent increase in mortality rates that hit Boston during the early 1870s, however, added the sense of urgency necessary to overcome resistance to increased public spending.[3] In an era still dominated by moral-environmentalism, improved sanitation seemed essential to mitigate the social effects of the 1873 depression.[4] Moreover, public projects promised to attract new businesses and provide jobs for the poor. In this spirit, Bostonians contemplated a huge park system and a new street plan for the business district in addition to comprehensive sewerage and water supply improvements.[5]

Sudbury Reservoirs

In 1856, Bostonians had enjoyed Cochituate water for only eight years, but the reservoir was already becoming inadequate. By 1860, the city was using water as fast as it flowed into the lake, leaving Boston vulnerable to drought. In 1870 and 1871, water in Lake Cochituate fell to dangerously low levels.[6] The Cochituate Water Board "enjoined [citizens] to

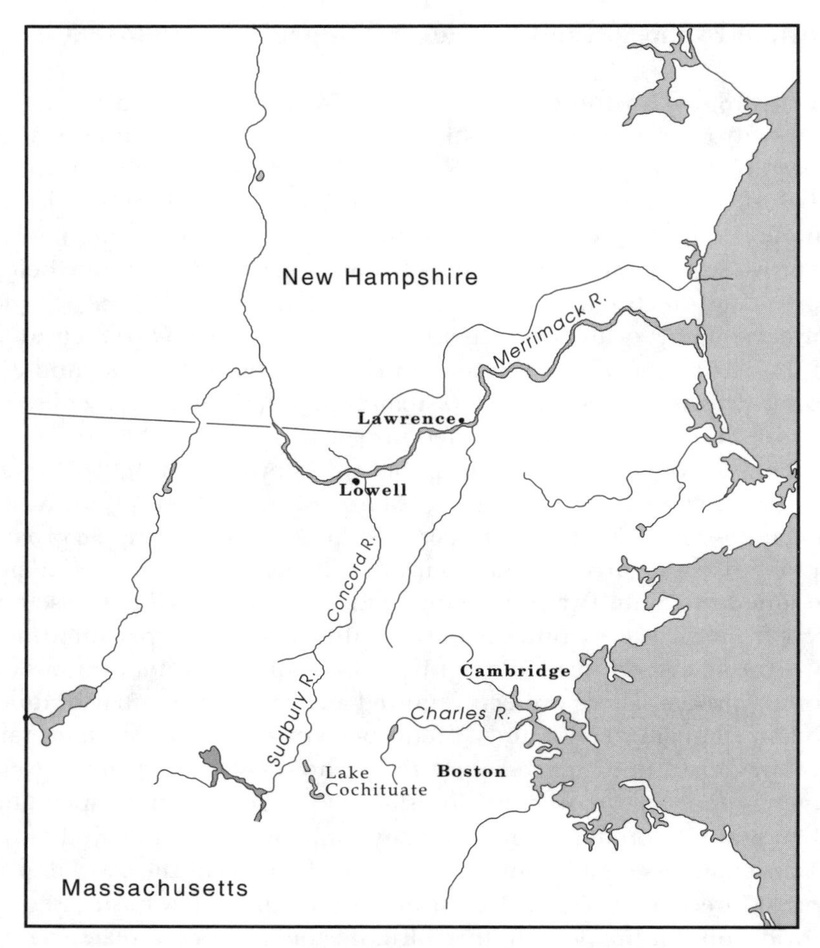

Sudbury River reservoirs. Beginning in 1872, water diverted from the Sudbury River supplemented Lake Cochituate supplies.

limit their consumption of water to the smallest possible amount."[7] Water conservation alone, however, could not solve the city's water problems for long. Moreover, reducing water use was seen as inconsistent with both economic expansion and improved public health.

Boston's water crisis of the 1870s had several major causes. In fact, the introduction of running water spurred increased water use by individuals and industries alike. The engineers who designed the Cochituate system had projected that Bostonians connected to the system would increase their per capita water use from three to five gallons a day to thirty gallons a day.[8] But the engineers underestimated; in most communities, running water brought with it per capita water use of between sixty and one hundred gallons a day.[9]

Furthermore, the demands on Cochituate increased as the city itself expanded. Boston's population grew from a population of 136,881 in 1850 to 250,526 in 1870 and 362,839 in 1880. Forty-five percent of that growth stemmed from annexations: Roxbury in 1868; Dorchester in 1869; and Charlestown, Brighton, and West Roxbury all in 1874.[10] Charlestown brought with it an independent water supply, but most of the other communities had their own water shortages. Indeed, in Roxbury access to Boston's water supplies had been a major argument in favor of annexation. By 1874, Boston found itself hard pressed to meet the commitments it had made to its new neighborhoods. Without adequate services, Boston had little hope of fulfilling the expansionist dreams of city leaders who wanted to annex Quincy, Chelsea, and Brookline as well.

Certain water policies increased the strain on Cochituate supplies as surely as annexation did. Shortly after completing the waterworks, the city council committed municipal funds to lay pipes to homes and businesses throughout the city as well as to build mains and branch lines.[11] This decision significantly reduced the costs to property owners of connecting interior plumbing to the waterworks. The city council hoped that this policy would give more Bostonians access to pure water, and so increase the sanitary benefits of the waterworks. By the 1870s, this policy had resulted in widespread, if not universal, use of the public water works, putting ever greater strain on the Cochituate supplies.

Under pressure to meet ever-increasing demands, the Cochituate Water Board began to look for new sources of supply in the early 1870s. In April 1872, following two particularly dry years, the board convinced the city council and the Massachusetts General Court to permit Boston to divert water from the Sudbury River into Cochituate when the reservoir dropped below acceptable levels.[12] The emergency nature of this measure did not satisfy those in Boston who sought a permanent increase in water supplies. Nevertheless, most Bostonians welcomed the completion of the temporary conduit between the Sudbury and Lake Cochituate in June 1872.

Any optimism regarding the future of Boston's water supplies was dashed four months later. On November 9, 1872, fire swept through

Boston's business district. The Great Fire spread unhindered for two major reasons. First, disease had weakened Boston's horse population; sick horses could not move fire equipment rapidly enough to contain the flames. Secondly, low water pressure, caused by low water levels in Cochituate and small water mains, prevented fire fighters from extinguishing the blaze.[13] The devastation inspired the city to install higher capacity water mains and hydrants in the burnt district. The disaster also seemed to justify permanent development of the Sudbury-Cochituate conduit.

In 1873, the Cochituate Water Board instructed city engineers to proceed with the surveys necessary to begin the conduit. Despite the sense of urgency fueled by the fire, debate over alternative supplies of water delayed final approval of Sudbury River development for two years.[14] These debates closely resembled the discussions of the Lake Cochituate project during the 1840s. Opponents feared higher taxes, cost overruns, and the costs of compensating water-mill owners for interference with their water power.[15] The *Boston Herald* maintained that the Sudbury's water was not suitable to drink, and insisted that the water board could find purer water at lower cost in Weston, Wellesley, or South Natick.[16] One engineer, quoted in the *Boston Globe,* raised fears of catastrophic failure of one or more of the Sudbury dams.[17]

A number of critics used the Sudbury proposals explicitly as an excuse to attack municipal administration of public works. The editors of the *Boston Post,* for example, complained that the city council was blind to impending hazards caused by water shortages. The *Post* was particularly concerned about fires, lost economic opportunities, and dangers to public health. The paper went on to advocate the creation of an independent water commission: "Were a water Commission established . . . every one would feel sure of prompt action, or what is much better, of ample provision . . . against an emergency like the present."[18]

Throughout the 1870s, a welter of such "emergencies" inspired many Bostonians to demand both new public works institutions and independent investigations into public service needs. Some historians have interpreted these appeals as evidence of the ethnic and partisan divisions that characterized late-nineteenth-century cities. According to their interpretations, Democrats generally supported traditional municipal governance, while Republican and Yankee reformers sought to circumvent their rivals by creating new governmental agencies. The "water Commission" proposed by the *Post* would have fit into this model of Gilded Age political reform, except that the *Post* was one of Boston's leading

Democratic newspapers. It was precisely the nonpartisan trust in special commissions and the equally general sense of urgency about maintaining effective services, such as water supply, that enabled Boston to build extensive public works first under municipal and later under regional supervision.

Meanwhile, the Sudbury project aroused opposition outside Boston city limits, as well. Beginning in 1872, Boston acquired control over more and more water in the Sudbury River. In 1874, armed with these new water rights, the Cochituate Water Board began building a series of dams on the Sudbury to increase the amount of water it diverted into Lake Cochituate. Although the General Court ordered Boston to provide water to Framingham, Newton, Brookline, and several other towns in compensation, the communities along the Sudbury naturally resented Boston's interference.[19] For the time being, however, they were powerless to block the city's expanding water claims.

By 1875, Boston had significantly increased its control over water resources in eastern Massachusetts. The Cochituate Water Board oversaw not only Lake Cochituate but also much of the Sudbury River. Annexation of Charlestown had integrated Mystic Lake into the Boston water system as well. Meanwhile, water supply contracts with Chelsea and Somerville, inherited from Charlestown, and the construction of waterworks to compensate towns in the Sudbury and Cochituate watersheds for lost water rights further increased Boston's power and responsibilities outside city limits.[20] Nevertheless, in less than a decade Boston went in search of yet more water. Unfortunately, this continued municipal activism seemed only to confirm the *Post*'s insinuation that city officials failed to protect Boston's interests.

The Main Drain

Although Boston struggled with fresh water shortages, it suffered from an overabundance of wastewater. Because an 1833 ordinance allowed Bostonians to connect privies to city sewers, the twentyfold increase in water use that created the need for the Sudbury reservoirs flooded Boston's sewers.[21] Decaying pipes, landfill, and low elevations created serious drainage problems. Meanwhile, sewage accumulated in shallow or stagnant waters along the waterfront, creating intolerable conditions in many neighborhoods.[22] In newly annexed but sewerless suburbs, wastewater overflowed privy vaults, saturating soils and threatening to create "a permanent evil . . . from a temporary convenience."[23] The municipal sewer department had spent fifty years installing new outfalls and dredging the waterways where sewers ended, all in response to complaints about these conditions. Between 1868 and 1875, the department redoubled its efforts, installing over seven miles of new pipes each year, but to little avail.[24] Because most Bostonians still feared filth diseases and still saw isolated unsanitary conditions as a threat to general health and moral order, local leaders had a strong incentive to undertake even more aggressive sewerage construction.

Three waterways, the Back Bay, the South Bay, and the Roxbury Canal, dominated sewage discussions in the 1870s. Enormous quantities of sewage accumulated in these basins; their odors spread across the city. In earlier decades, tidal currents had scoured all three waterways, preventing serious sewage nuisances. By the 1870s, landfill and urban development had left them shrunken and nearly stagnant. Meanwhile, sewer lines extending through the fill ran nearly level, with outfalls below the high tide line. Thus sewage moved extremely slowly, allowing debris to accumulate and rot in the sewers. Even worse, incoming tides swept sewage backwards up the pipes. During storms, this backward flow could carry sewage into basements and onto city streets.

The Back Bay attracted municipal attention first, partly because problems there were quite severe, but also because this neighborhood was wealthier than either the South Bay or the Roxbury Canal area. In 1850, the city council hired Ellis Chesbrough and William P. Parrott to help devise a solution to Back Bay drainage problems. Chesbrough and Parrott noted that tidal waters flowed into sewers at high tide and that Boston had established no consistent depths or slopes for sewers built by private parties. To solve the first problem, they recommended installing tidegates to keep ocean water out of the sewers. For the second, they suggested that the city council establish standard grades and elevations for all city sewers. Such standards would ensure that all sewers flowed along gradual, even slopes to the sea.[25]

Ten years after Chesbrough and Parrott's report, the sewer department found that sewer grades in the Back Bay were still inadequate for proper drainage. In 1861, the city built a new main through the neighborhood, all the while lamenting that the worst branch lines lay too far below the street to be connected to it. By 1867, conditions in the neighborhood had so deteriorated that the sewer department supported a proposal to improve drainage by raising the elevation of the entire Back Bay

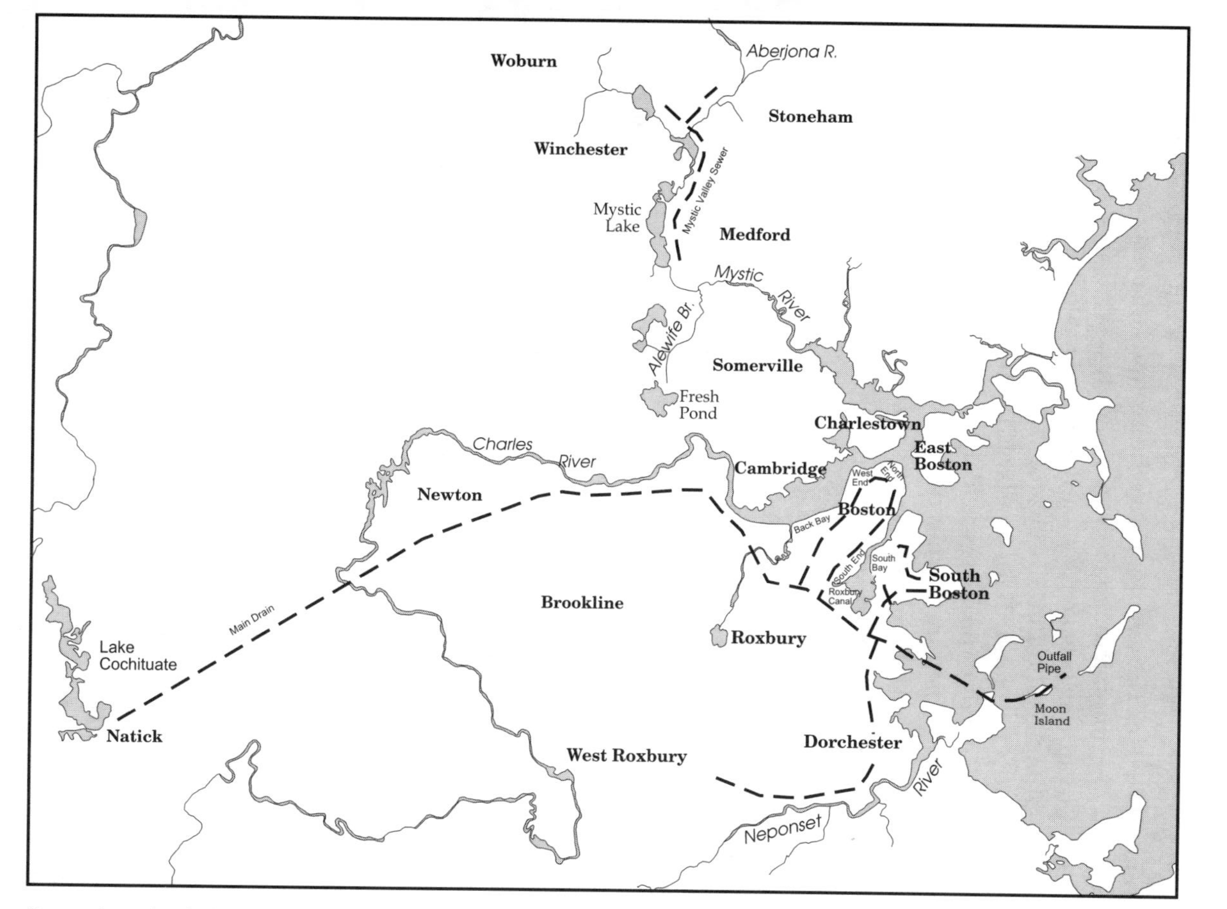

Boston's main drain and Mystic Valley sewer. Completed in 1884, the main drain was intended to eliminate sewerage nuisances in the Back Bay and South End. The construction of the Mystic Valley sewer caused bitter conflicts between Boston and Medford.

neighborhood.[26] By this time, business leaders had begun to pressure the city to improve Back Bay conditions on the grounds that poor sanitation contributed to economic stagnation as well as disease.[27] Economic arguments were to appear repeatedly over the next decade and to carry significant weight with city administrations.

In the South Bay, landfill for roads and commercial development caused the same kinds of problems as had residential fill in the Back Bay. Odors, poor drainage, and frequent flooding of streets and cellars plagued the neighborhood. In the late 1860s, for example, a new waterfront roadway, Atlantic Avenue, cut off six large sewers. Although Atlantic Avenue developers diverted the pipes to new outfalls, one resident demanded that the city council "abate the intolerable stench that the residents of the South End are subject to almost nightly."[28] Regrettably, the effort to remedy the Atlantic Avenue problem was typical of Boston's midcentury approach to sewerage: mitigating a specific nuisance in isolation from other drainage problems.

South End residents endured stenches from both the Roxbury Canal and the South Bay. The canal, dug in the 1790s, had originally permitted ships to carry cargo to Roxbury wharves at low tide.[29] By the 1860s, however, the canal functioned primarily as a cesspool for industries along its shores. In the 1870s, the Boston Board of Health blamed "foul gases from the putrid bottom" of the canal for high mortality rates in the South End.[30] A city council sewer committee acknowledged in 1873 that a variety of industries, including soap factories, were causing serious problems. In response to committee recommendations, the city council dredged the canal to remove offensive materials, and tried to divert industrial wastes around both the canal and South Bay. Neither of these strategies improved conditions for long. Within a few years, the city council came under renewed pressure from prominent residents and physicians in the South End.[31]

Not only did persistent sewerage nuisances inspire complaints from all parts of the city, and not only did Boston's improvement projects utterly fail to reduce the numbers of petitions citing these problems, but increasingly the Boston city council found itself under fire from local health officials. Boston's health officials were among the first to call for citywide sewerage improvements.[32] In 1874, the Boston Board of Health—still subscribing to environmental explanations of infection—warned that "prevalent summer diseases are influenced by [the] poisoned atmosphere" caused by sewage pollution.[33] Because typhoid, diphtheria, meningitis, cholera, and pneumonia caused the majority of preventable disease in the city, prevention of these filth diseases constituted "without any doubt the greatest and most urgent sanitary need of Boston."[34] The continued emphasis on the relationship between filth and disease transformed unpleasant odors and unsanitary conditions into public concerns of the first order.

A 1875 Boston Board of Health report foreshadowed a transition from a purely environmental approach to public health to the technological and bacteriological strategies that would come to dominate sewerage planning. Now the Board explained that "microscopic forms, apparently of the lowest vegetable life" caused infectious disease in the city. But it also noted that "filthiness of the air, water, and soil . . . may be a more or less necessary condition for the evolution of the specific virus to which the disease owes its origin. . . ."[35] Thus, while acknowledging the latest in scientific discovery, the report still recommended the "rapid and continuous translation of our sewage . . . [to] the terminal outlets of our public sewers" as the best way to prevent disease.[36]

In addition to public health, some advocates of improved sewerage emphasized the economic benefits of good sanitation. They argued that cleaner streets would increase property values and that modern infrastructure would attract business. North End petitioners blended public health and economics in an 1864 petition to the city council declaring that better sanitation would increase land values as well as improve the health of local residents.[37] Sometimes, appeals to public health concealed self-interest. Such was the case with a proposal for a Back Bay park, sent to the city council early in Boston's reevaluation of city sanitation. Back Bay investors proposed that the city build a park with a basin designed to flush out the neighborhood's sewers.[38] The park would have taken unusable acreage off investors' hands while making remaining lands more valuable and inhabitable. Instead of calling attention to these benefits, Back Bay investors claimed that the park would draw fresh air into the city and thus improve general public health. The Back Bay park received substantial support for both its stated and unstated goals. To pass muster, economic self-interest had to be cloaked in the rhetoric of public benefits but, at the same time, costly public improvements required a measure of economic justification.

In 1875, the Boston City council appointed a board of two nationally recognized consulting engineers, Ellis S. Chesbrough and Moses Lane, and a well-known sanitarian, Charles F. Folsom, to devise a solution to the persistent sewage problems in the Back Bay, the South Bay, and Roxbury Canal, as well as the Charles River, Stony Brook, and the waters around Dorchester. The city council's instructions to this sewer commission specifically directed Chesbrough, Lane, and Folsom to evaluate proposals for a "high-water basin" that could grace a Back Bay park and flush clogged sewers in its vicinity.[39] Despite this concession to powerful Back Bay investors, the city council's instructions and choice of personnel revealed growing impatience with the piecemeal construction that had characterized Boston's sanitation for decades. As their records would have made plain to the city council, Chesbrough, Lane, and Folsom were not inclined to examine specific nuisances in isolation. Unlike the Boston sewer department, they recognized the inexorable connections between Boston's drainage problems and conditions throughout three entire river basins.

Chesbrough's credentials were, by this time, well known to most Boston leaders. He had spent several years as chief engineer for the Cochituate Water Works, consulted on Back Bay sanitation in the 1850s, and earned a national reputation for his work in Chicago.[40] Lane, an associate of Chesbrough's, had designed and constructed water systems in Pittsburgh, Indianapolis, and a number of other cities. Folsom had earned his reputation as a sanitarian in New England, and served on the Massachusetts State Board of Health before working on the Boston sewer improvements.[41] His inclusion on the board reflected a continued preoccupation with the political and social aspects of public health. Subsequent consulting committees would include only engineers, which reflected a national trend to view public services in purely technical terms.

In 1876, Chesbrough, Lane, and Folsom issued their report. They identified many of the same sanitation problems recorded earlier: leaky cesspools and pipes, tide-locked sewers, and poorly constructed pipes clogged with years of accumulated sediments. They estimated that every day Boston's thirty-two fully independent drainage networks and forty sewer outfalls delivered some twenty million gallons of sewage into Boston's harbor and rivers. Because nearly all the pipes opened between docks or into shallows, wastes were frequently exposed at low tide, and quickly washed ashore. Earlier construction, they charged, had failed to improve sanitation because of poor planning and inadequate engineering that had resulted in inefficient drainage.[42]

To solve these problems, Chesbrough, Lane, and Folsom proposed a sewage disposal system that came to be known as the Main Drain. They recommended that Boston build intercepting sewers along the waterfront to collect sewage before it spewed into the harbor. These pipes, large enough to hold up to seventy-five gallons of waste per capita, would transport wastes to holding tanks on Moon Island. From there, the sewage would flow into the harbor only during the first two hours of each outgoing tide. When combined rain and waste waters exceeded pipe capacities, overflow outlets would send excess sewage through the old outlets.[43] This proposed system would allow sewage to drain out of the city regardless of tide and storm conditions and, except during severe rainstorms, would conduct all the city's sewage away from its waterfront.

Significantly, Chesbrough and his colleagues envisioned interceptors along both shores of the Charles River. Their report called for a northern sewer to drain Cambridge, Somerville, Everett, Chelsea, Revere, and Winthrop. Many details of this part of their plan were subsequently incorporated into the regional system.[44] At the time, however, constructing this segment of Chesbrough's plan, which would have almost doubled the cost of the Main Drain, would have required extensive cooperation from northern suburbs. It was unlikely that the communities north of Boston would have embraced such a huge symbiotic project. And, for both political and financial reasons, Boston officials shrunk from the thought of paying to sewer the towns north of the Charles River. In vetoing this element of Chesbrough's plan, Boston's city council rejected an early bid for a regional approach to public sanitation, expressing instead a naive faith that Boston could fix its sanitary problems on its own.[45]

Although the city council quickly disposed of proposals that Boston build sewers north of the Charles, other elements of the Main Drain plan occupied their attention for much of 1876. A few councilors objected to Chesbrough's plan because they thought the city should treat or reclaim valuable nutrients from its sewage. Some insisted that the city ought to investigate lower cost alternatives to the interceptors and harbor outfall. Others objected to the influence of the wealthy on Boston's public health projects. Invariably, sewerage supporters countered with references to the public health needs of the city and the social devastation they expected if sewage pollution continued.

The consultants recommended that Boston simply dump its wastes into the harbor because they believed that the ocean could purify large quantities of sewage. Moreover, this method of disposal cost much less than the known alternatives. Although a few city councilors decried harbor disposal as wasteful, long before the 1870s the sale of nightsoil had ceased to generate profits.[46] Moreover, running water diluted domestic wastes to the point that they had little value as fertilizer. Furthermore, Boston's size and soil conditions ruled out the use of waste water for irrigation, and few other effective treatment options existed at this time.[47]

Only one city councilor, George L. Thorndike of East Boston, opposed ocean disposal because he believed it would contaminate the harbor. He feared that wastes would accumulate among the islands and "sooner or later . . . cause trouble." This opinion revealed rare prescience. Thorndike attracted little support, however, because at this time, most people still believed that moving water could purify itself.[48] In Chesbrough, Lane, and Folsom's defense, deep water disposal did eliminate the odors so feared in the nineteenth century. Furthermore, for those who accepted germ theory, the Moon Island outfall appeared distant enough to prevent human contact with infectious organisms. Thus the sewage disposal provisions of the Main Drain reflected current understanding of both the dangers of accumulated wastes and disease etiology, and balanced this knowledge against the cost of infrastructure.

Aside from Thorndike's objection, the only other element in the 1876 sewerage report that received extensive criticism was the recommendation that Boston continue to use a single set of pipes for both household waste and storm drainage. A number of experts, most notably George Waring, insisted that separating storm and domestic drainage provided better sanitation for less money than did combined systems. In Boston, Superintendent of Sewers William Bradley and West End councilors Warren K. Blodgett and Patrick F. McGaragle spoke in favor of separate sewerage. They saw inadequate storm drainage in low lying neighborhoods as Boston's main sanitary problem. Isolating disease-bearing household wastes from rainwater, they thought, might make expensive, comprehensive sewerage unnecessary.[49] At this time, however, no large city had attempted to build a separate sewer system. Bradley, Blodgett, and McGaragle attracted fewer supporters than they might have, had Waring completed his model projects in Lenox, Massachusetts and Memphis, Tennessee when Bostonians debated this issue. Moreover, because Boston already had a combined system, building a separate system would have required completely resewering the city, which, whatever the impact on operations, would clearly have cost more in the short run.

In August 1877, the Boston city council voted resoundingly in favor of the Main Drain as outlined in Chesbrough's 1876 sewer report. The Common Council approved the measure fifty-two to five, and the Aldermen nine to one.[50] As Alderman Choate Burnham noted, "The public have accepted the belief, as one needing no further evidence, . . . that our sanitary condition is so bad that it must be improved by a radical change in our system of sewerage."[51] Several councilors from the Back Bay, still hoping to get their park added to the plan, withheld their support. But the promises of quality sewerage service, economic growth, and a reduction in vice and disease were definitely attractive to representatives from the rest of Boston's wealthy wards. Meanwhile, the promise of construction jobs for the unemployed appealed to representatives of poorer neighborhoods.

As with the regional public works to follow, the Main Drain overcame partisan conflicts because the project itself addressed a set of conditions that affected a broad cross section of Boston's population and because it gave almost every major interest represented in Boston's municipal government something that it wanted.[52] Only holdouts for the Back Bay park were disappointed. But the Boston city council of the 1870s appeared unwilling to undertake expensive projects for the benefit of a few propertied tax payers. By the time of the Main Drain, Boston leaders appear to have abandoned essentially nondemocratic public works in favor of broadly distributed public benefits.[53] This transition was all the more important because the council's new attention to the whole city pushed Boston to build comprehensive rather than piecemeal public works.

Boston completed the Main Drain without much fanfare in 1884. All too soon, however, it became clear that the Main Drain alone was insufficient. As the Chesbrough report had suggested, in the absence of a similar project along the north bank of the Charles River, sewage borne past Boston from upriver communities continued to plague much of the city. Unfortunately, a regional sewer remained beyond Boston's capabilities. This was a case in which the failure of a city's leaders lay not in corruption or lack of will, but in institutional and territorial constraints on municipal power.

Mystic Valley

By the time the Main Drain was completed, Boston residents had learned much about the limitations of municipal public works. The Sudbury Reservoirs demonstrated that competition for water resources and for public works investment could interfere with critical projects. The Main Drain left Bostonians dissatisfied because they had little influence over upstream waste disposal. Thus, even though Boston had acquired considerable powers outside city limits, its leaders were still unable to provide adequate water and sewerage for the city.

Nowhere was the combination of territorial constraints, interlocal rivalry, water pollution, and inadequate water supply more problematic than in the Mystic River Valley. The northernmost of the four rivers in the Boston metropolitan area, the Mystic flows only twelve miles from its headwaters near Woburn to Boston Harbor. From 1868, Charlestown supplied Chelsea and Somerville with water from Mystic Lake under an interlocal contract. A dam below Mystic Lower Pond separated the upper lake's fresh water from the tidal lower Mystic River. By 1874, the Mystic was providing drinking water to four cities and draining wastes from several other towns and their industries. Boston and the Mystic Valley towns came into conflict precisely because of these mutually exclusive uses of the river. Although the Mystic Valley quandary never received as much press as Back Bay pollution or Lake Cochituate water shortages, the Mystic provided powerful evidence of the need for regionalism.

In the 1870s, twenty-three tanneries, a rendering plant, and a felting mill spewed an unappetizing mix of lye, organic matter, chlorine, and other chemicals into the Mystic. Medford, Stoneham, Winchester, Woburn, and other riverside towns added their sewage and street drainage. Waste disposal in the Mystic was neither malicious nor unusual. Americans had long dumped wastes into the nearest watercourse, believing that rivers could cleanse themselves. Moreover, the practice cost less and worked as well if not better than existing alternatives. Boston had so thoroughly embraced water disposal for sewage that the sewer department had to convert several of the city's smaller streams into sewer mains.[54] By the late 1860s, however, Charlestown residents were complaining that upstream pollution gave Mystic water an unpleasant odor and color. The wastes also damaged steam engines and industrial machinery. Such thick layers of sediment accumulated in boilers that, by some estimates, pollution in the Mystic reduced the useful life of machinery by as much as ninety percent. By the mid-1870s, fish were dying and cattle were refusing to drink from the river.[55] Boston needed the Mystic, however. Even with the Sudbury additions, Cochituate did not have enough water for East Boston and Charlestown in addition to the rest of the city. Boston water officials had no choice but to try to keep contaminants out of the Mystic River.

In May, 1874, the Boston city council's water committee recommended building a sewer to "convey away the impurities discharged into

Mystic Pond." Meanwhile, Boston's Mystic Water Board suggested a symbiotic project with the Mystic Valley towns to divert wastes around the lake. Boston leaders expected cooperation because state officials had begun to pressure the Mystic towns to improve water quality. The State Board of Health, in 1873, had prohibited Medford from building municipal sewerage that would have increased river contamination. The state supreme court had also ruled against further contamination of Mystic Lake.[56] By the time Boston petitioned the legislature for permission to build the sewer, the Mystic towns, resentful of Boston's increasing influence and unwilling to bear the costs of their part of the construction, refused to cooperate.[57] These towns' on again off again endorsement of comprehensive sewerage made Boston extremely wary of interlocal cooperation in the Mystic Valley. Without their support, however, Boston would not build a sewer large enough to carry their domestic wastes.

In spite of Mystic Valley opposition, the Massachusetts General Court granted Boston the authority to "take, hold, and convey . . . any or all the water belonging to the watershed of the valley of the Mystic . . . which flows . . . directly or indirectly, into Mystic Pond."[58] This legislation permitted Boston to build sewer and water networks without consulting other communities and to include or exclude anyone it wished from its water or sewer systems. Boston could now outlaw any shoreline activities that might pollute its water supply. The Mystic towns saw their future placed in Boston's hands. Although intended to mediate between two mutually exclusive uses of the Mystic River—water supply and waste disposal—the 1874 Mystic Valley Sewer legislation gave Boston nearly exclusive control over water resources and, by extension, commercial and domestic development along the Mystic River.

Boston now seemed ready to exercise its authority over the Mystic, but no action was immediately forthcoming. In November 1874, the Mystic Water Board proposed a small $160,000 sewer to carry industrial wastes around the lakes. The Board acknowledged that this sewer would provide no protection from domestic sewage, particularly if the Mystic towns built sewer lines to carry their domestic wastes to the river. Additionally, several city councilors, including George P. Denny of the South End and Uriel H. Crocker of Back Bay, argued that the sewer would not significantly improve Mystic water quality and thus was a waste of the city's money.[59] Eugene Sampson, too, strongly opposed the industrial sewer, insisting that the Water Board proposed it only "in the hope of forcing the [Mystic Valley] towns to join in building a large sewer sufficient to drain that valley."[60]

In the end, financial considerations played a major role in Boston's decision to build an industrial rather than comprehensive drain. Although even the most ambitious Mystic plan would have cost only $600,000, a mere tenth of the Main Drain's cost, city councilors had few incentives to spend money to provide sewer services to residents of the Mystic Valley.[61] Moreover, since the Mystic water system served some of Boston's least influential neighborhoods, an expensive project was not politically appealing. Finally, the Mystic Valley Sewer ran afoul of the same kind of objections that defeated the Back Bay park during the Main Drain debate. Many wealthy Bostonians maintained summer residences in Medford; if Boston had built a comprehensive sewer without financial contributions from the Mystic towns, the city council would have appeared to be spending general funds for the benefit of a privileged few.[62]

In 1875, after a year of debate, Boston was no closer to eliminating pollution in Mystic Lake. The city council made one more bid for comprehensive sewerage, petitioning the General Court for authority to build an elaborate, symbiotic project. Although the legislature quickly approved the necessary special legislation, comprehensive sewerage in the Mystic Valley remained politically unrealistic. By the end of the legislative session of 1875, cooperation between Boston and the Mystic towns again collapsed, apparently confirming Bostonians' suspicions that these communities were "bound to make the city of Boston furnish drainage for the whole Mystic Valley."[63] Boston certainly would not pay the whole cost of sewerage under these circumstances and finally abandoned hopes for a large drain in the valley.

In 1876, the Boston city council approved a small-scale sewer intended to divert only industrial wastes around Mystic Lake, as recommended by the Mystic Water Board in 1874. As East Boston and Charlestown residents discovered as soon as the Mystic Valley Sewer was completed, the limited goals for the sewer hurt Boston as much as they did Medford and its northern neighbors. Natural drainage and municipal runoff continued to find their way into the river, so pollution of Mystic Lake continued in spite of the Mystic Valley Sewer. Friction between Boston and the Mystic towns deepened when Medford found that Boston's project stymied its plans for a town sanitation system. The fear that Boston would control development in the valley proved even harder to swallow. The campaign for the sewer left a residue of resentment, while the sewer itself limited the Mystic towns' political and economic autonomy.

Four short years after the Boston city council endorsed the Mystic Valley Sewer, Medford had the opportunity to fight back. In late November 1880, parts of Mystic Lower Pond froze in a sudden cold snap. Waste from the Mystic Valley Sewer poured onto the ice, and "the whole town . . . was aroused by a stench that almost took one's breath." Noxious gases discolored houses near the lake shore and killed fish. Eels crawled out of the pond onto frozen ground to escape the foul water.[64] Armed with lurid descriptions of these conditions, in early 1881, Medford petitioned the General Court to close the Mystic Valley Sewer. Conditions in the lower pond ultimately forced state and local officials to reexamine comprehensive waste disposal in the greater Boston area.

The events of 1880 should have surprised no one. In 1875, Arlington residents had opposed the sewer precisely because it "would convert the Pond into an offensive cesspool, destroy the fish therein, create a common nuisance, and cause great injury to the public health."[65] Despite these predictions, Boston felt it had no viable alternative to the lower pond outfall. The State Board of Health prohibited Boston from discharging sewage into the river below the lower pond, ostensibly to protect the lower reaches of the river, but also to encourage the city to consider a more comprehensive project.[66] The Board advocated a harbor outfall, but Boston saw that as the first step towards providing the Mystic valley with its own Main Drain, on the backs of Boston taxpayers.[67]

In January 1881, Medford representative John C. Rand introduced a bill in the Massachusetts House of Representatives to "require Boston to abate a nuisance caused by it in Mystic Lower Pond."[68] Medford and Arlington petitioned the Committee on Water Supply and Drainage to repeal the 1875 Mystic Valley Sewer Act and require Boston to clean the pond. The House referred the matter to the Committee on Public Health, which Rand chaired. Three months later, the committee recommended that the House pass Rand's bill. Only one of Boston's three representatives on the committee, Henry Lyons, dissented. He contended that the factories connected to the Mystic Valley Sewer should assist in cleaning up the pond. The General Court placed responsibility for the sewage nuisance squarely on Boston's shoulders, however.[69]

Rand's bill passed on May 13. In the name of public health, it required Boston to abate the nuisance in the pond and shut down the Mystic Valley Sewer. The legislation allowed Boston to reopen the sewer only if it "purified, cleansed and freed the said waters from all offensive, contaminating, noxious and polluting properties and substances."[70] This victory must have gratified longtime opponents of the Mystic Valley Sewer, but it did not immediately improve the Mystic Valley's sewerage.[71]

The nuisance in Mystic Lower Pond reopened the question of regional sewerage by demonstrating that cities could not exercise enough control over external factors to provide adequate services to their citizens. Like cities along many American rivers, the communities involved in the Mystic dispute sought the least expensive means to provide themselves with water supply and waste disposal. As a result, each town used a single waterway for both services and left the towns downstream to fend for themselves. This arrangement was adequate for isolated agricultural communities, but proved unacceptable in industrial, densely populated areas. In the Mystic Valley, eliminating pollution required removing both industrial and domestic wastes. But in the 1870s, Boston and the Mystic towns remained unwilling to undertake such an ambitious project. Boston voters, while content with expensive projects that clearly benefited themselves, would not tolerate similar expenditures on behalf of other towns. Moreover, because any external sewer inevitably expanded Boston's economic and political reach, even a sewer that met all of Medford's or Arlington's requirements would have increased interlocal tensions in the Mystic Valley. Boston and the Mystic towns faced a quandary for which existing institutions offered no good solution.

In the aftermath of the Mystic Lower Pond incident, the Commonwealth of Massachusetts undertook a series of water resources investigations in an attempt to ascertain the water supply and sewerage potential of the state's most heavily used or contested rivers, including both the Mystic and the Charles. These studies began with the assumption that Massachusetts communities could not provide themselves with services without state mediation and centralized planning. Boston's public works struggles, particularly the Sudbury, Main Drain, and Mystic projects, were essential to bringing state and local leaders to this conclusion.

The Mystic, Main Drain, and Sudbury projects clearly demonstrated both the political and the environmental limitations of municipal enterprise and thus were crucial to the transition from municipal to regional systems. Local politics, interlocal conflicts, and financial barriers proved insurmountable obstacles when a city or town attempted to solve its water and sewerage problems by itself. Recognition of the failure of municipal enterprise and impatience with the limited success of interlocal cooperation finally combined to push Boston and its surrounding communities towards regionalism.

East Bay: Public Services Frustrated, 1900–1920

Oakland confronted the limitations of municipal power shortly after undertaking its first public works projects. Interlocal rivalries and political borders limited municipal and cooperative enterprises much as they had in Boston. Furthermore, the East Bay faced several additional barriers to the expansion of municipal authority. Private corporations still owned not only the East Bay's water systems, but also all available sources of drinking water. Moreover, because water service was considered to be a utility rather than a public health necessity, the East Bay cities had no clear mandate to extend their authority over waterworks. Nevertheless, East Bay officials reacted to constituent complaints with strategies remarkably similar to those used in Boston.

As described in the previous chapter, by the turn of the century Oakland had lost a number of battles to establish public waterworks. Nevertheless, the city had expanded its authority to regulate private utilities, including the water companies. Municipal responsibility for sewerage remained unchallenged, although Oakland struggled to control flooding that originated outside the city. Meanwhile, population growth was creating incentives to improve services in the East Bay. For example, many San Franciscans relocated in the East Bay after the 1906 earthquake and fire destroyed their homes; this surge in population placed new demands on existing infrastructure. In addition, East Bay leaders saw improved services as the key to greater prosperity. Economic motivations to provide services grew even stronger as water shortages began to increase the costs of doing business in the East Bay.

Significantly, public health was not a major consideration in either water or sewer debates. By 1900, the germ theory had supplanted moral-environmentalism in American sanitary planning, reducing the problem to a matter of separating people from infectious wastes. With the triumph of germ theory, expectations that water supply and sewerage would transform society faded, as did the urgency and popular appeal of many public works proposals. As a result, East Bay campaigns frequently devolved into explicit discussions about the ambitions and corruption of local leaders. Conspicuously missing were the appeals to altruistic reform so important to Bostonians. Unfortunately, the promise of commercial growth and political reform alone were insufficient to create and sustain the political consensus necessary to unseat private water companies or induce East Bay communities to undertake truly innovative public water and sanitation projects.

In their quest for public ownership of waterworks, the East Bay cities experimented with a variety of new institutions. During this period, Richmond voters approved the Richmond Water Commission, the East Bay's first municipal water agency. Oakland gave ambivalent support to San Francisco's Hetch Hetchy project and considered purchasing water rights for an independent municipal system. None of these proposals seriously threatened local water companies. They did, however, demonstrate an interest in interlocal cooperation and in institutional innovation that closely resembled Boston's efforts to expand municipal authority. Meanwhile, the East Bay's private water companies continued to expand their control of local water supplies, despite complaints about water shortages and high prices. As the largest, richest, and most powerful city in the East Bay, Oakland was in the best position to challenge private utilities. But Oakland's efforts to establish a public waterworks fared no better than its neighbors'. As in Massachusetts, interlocal conflict and overextension of existing networks bred frustration with and contempt for municipal enterprise. Hindered by these common problems as well as by their own limited authority over powerful utility companies, East Bay cities appeared incapable of building the enormous projects that their residents needed.

East Bay Sewers

Early-twentieth-century East Bay municipal sewer policies clearly reflected the influence of bacteriology. As fear of germs replaced fear of miasma, sanitary goals shifted and narrowed. East Bay sewer departments abandoned the comprehensive and aggressive approaches to waste disposal that had marked not only Boston's Main Drain, but also Oakland's earlier approach to drainage. Instead, they built systems that merely separated people from their wastes. Where necessary, they also targeted specific drainage and flooding problems. As a result, East Bay sewer projects during the first two decades of the twentieth century followed the pattern of small scale construction in response to immediate crises that had characterized municipal sanitation a hundred years earlier.[72]

In returning to smaller scale construction, Oakland in particular found itself reacting to one localized crisis after another. Piecemeal construction might remedy a single problem, but did little to prevent new ones. Oakland annexed over forty-three square miles of land during this period. This territorial expansion, combined with rapid population growth significantly increased the difficulties and costs of sewering the city.[73] Oakland was spared the immediate implications of its shortsighted approach to sewerage because the East Bay had naturally good drainage and because decades earlier, the sewer department had installed very large sewer mains. Thus, the first sewerage crisis to emerge resulted from poor disposal practices rather than inadequate sewer pipes.

Small-scale construction appealed to Oaklanders because it cost so much less than comprehensive sewerage. As its late-nineteenth-century efforts to improve water quality in Lake Merritt demonstrate, pressure to maintain low taxes while meeting service responsibilities in the growing city pushed Oakland to abandon the aggressive approach it had taken in earlier decades. By the 1890s, sewage from much of eastern Oakland had started to interfere with recreational use of the lake. Oakland officials directed consulting engineers Marsden Manson and C. E. Grunsky, noted for their work on San Francisco's sewers, to find an inexpensive solution to Lake Merritt pollution. Manson and Grunsky recommended that Oakland continue to permit storm drainage into the lake, but divert domestic wastes around the lake and into the Oakland estuary via separate pipes. They proposed that Oakland dig a short tidal channel and use it for waste disposal "until such time as it becomes offensive." Only "when the quantity of sewage delivered into the tidal channel becomes so great as to pollute the waters thereof," should the city begin pumping sewage into an offshore outfall.[74] This strategy was acceptable because sewage was now seen as unpleasant or inconvenient, and not as a life-threatening hazard.

Inevitably, Oakland's cost-cutting approach to Lake Merritt did not protect the lake for long. In 1902, Oakland had to make further investments to reduce lake pollution. Elsewhere in Oakland and its environs, additional water pollution and drainage problems emerged that resulted from excessive thrift, and which occasionally embroiled Oakland in interlocal conflicts rather like those seen in Boston.

Drainage from the hilltop towns of Elmhurst and Piedmont was one of the problems that created tension between Oakland and its neighbors. In 1924, to reduce flooding inside its own boundaries, Oakland scrambled to improve drainage in a small creek in neighboring Elmhurst.

Enraged by damage caused by creek flooding, Elmhurst residents began filling the creek and chasing away Oakland sewer workers.[75] Oakland finally appeased Elmhurst residents by cleaning the creek rather than constructing a culvert across private land.[76] A similar conflict erupted in 1927, when Piedmont's storm sewers overflowed Oakland's Grand Avenue sewer mains. "Grand Avenue's storm sewers," the Oakland Sanitary Commission declared, "cannot possibly care for the enormous storm drainage of the . . . city of Piedmont." Oakland tried to force Piedmont to make the necessary sewerage improvements, but Piedmont refused on the grounds that a 1925 contribution of sixty-five thousand dollars towards joint construction should have covered any subsequent improvements.[77] In these battles, Elmhurst and Piedmont residents wrung concessions from Oakland. Their ability to resist Oakland's will reflected their wealth and their influence in the Oakland business community. These suburban victories ensured that interlocal construction was neither as one-sided nor as divisive in the East Bay as in Boston. These sewer battles did, however, prevent Oakland from fully demonstrating municipal authority within and beyond city limits.

The shortcomings of Oakland's piecemeal sewerage eventually caught up with the city, but because of voters' aversion to increased public expenditures, Oakland leaders moved with characteristic slowness to remedy new sewage problems. In his 1928 annual message, Oakland Mayor John L. Davie announced that "one improvement absolutely mandatory is that of the city's main drainage and sewer system." He argued that immediate steps were needed to relieve the severely overburdened the network, but none were taken.[78] In 1933, Oakland city engineer Walter N. Frickstad recommended that the city spend six to ten million dollars to build a "perfect" system of drainage. At the very least, according to Frickstad, Oakland needed to appropriate one million dollars for urgent repairs for three sewer mains.[79]

In spite of Davie's and Frickstad's recommendations, years passed before Oakland seriously reconsidered piecemeal construction. Waste collection and disposal issues attracted attention in individual neighborhoods, not from the city as a whole. Thus officials had little incentive to make sewerage a priority. Once, fear of epidemics and social chaos had overcome voters' desire to minimize public expenditures. But by the time Oakland confronted widespread sewer nuisances, the germ theory undermined the connection between sanitation and public health that had created a sense of universal urgency in Boston. Therefore, Oaklanders voted with their pocketbooks and let sewerage development languish. Only when nauseating fumes from sewerage pollution of shoreline waters threatened economic development, as happened in the late 1930s, did Oakland officials consider comprehensive drainage.[80] Then, residents held municipal leaders responsible for letting conditions decline so deplorably.

Municipal authority over sewers and other public services lagged in the East Bay for many reasons. Without moral-environmental theory, local sanitary nuisances could be treated as mere inconveniences rather than citywide crises. In an era of fiscal retrenchment, East Bay voters and leaders alike were committed to low taxes and limited public authority. Entrepreneurs could and did provide many services judged not essential for public development, but depending upon private capital in this way did not help cities like Oakland to secure nonremunerative services such as sewerage. These barriers were sufficient to delay significant sewer construction until the 1940s, but advocates of increased public authority over waterworks encountered even greater obstacles.

Bay Cities Water Company

Although East Bay residents had originally welcomed private water development as an ideal means to improve services without raising taxes, by the turn of the century, their enthusiasm waned. High water rates and failing water supplies inspired considerable grumbling. Proponents of government services saw this growing dissatisfaction with the status quo as an opportunity to act upon both Progressives' criticisms of corporate monopoly and Californians' fear that utilities corrupted state and local politics. Nevertheless, East Bay public water proposals in the early twentieth century met the same fate as their antecedents. Public complaints could overcome neither the power of the private utilities nor their monopoly on water resources in the East Bay.

In 1905, the Bay Cities Water Company offered to sell Oakland its rights to Alameda Creek for use in a municipal system. This Bay Cities proposal offered Oakland what it needed most in order to build a publicly owned water system: water rights. Oakland seemed ready for public water development. The Bay Cities offer came only a few years after the Oakland and Contra Costa Water Companies merged. The merger inspired in Oakland residents fear of increased power for the corporate monopoly and higher water rates for local residents. Water was running

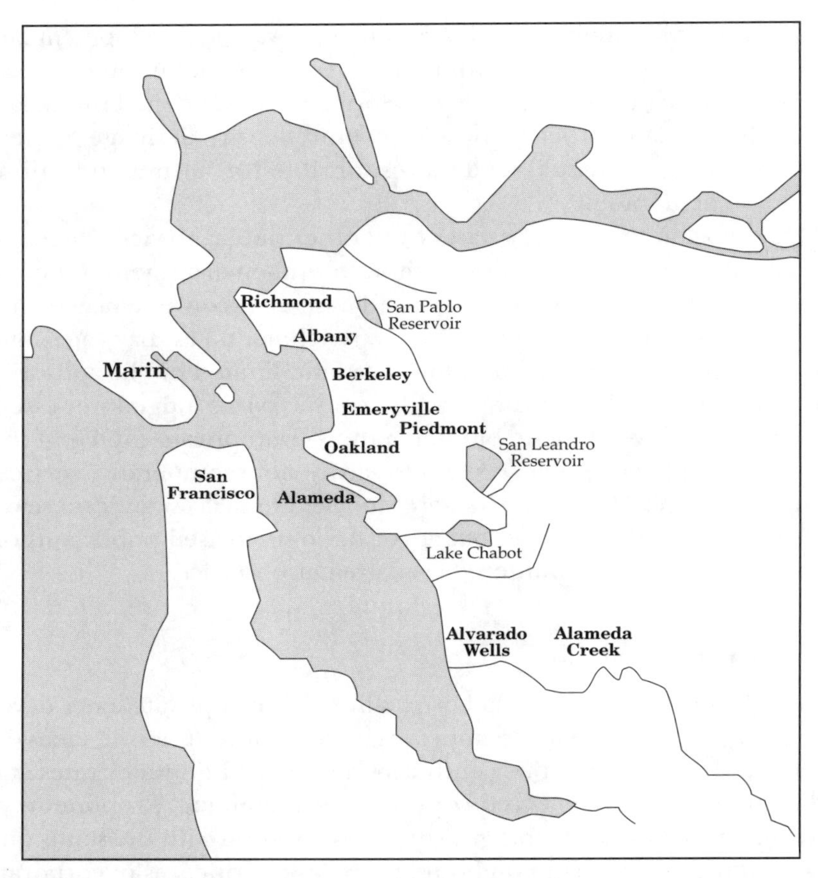

East Bay water supplies. By 1905, all East Bay water supplies were in private hands. Oakland briefly considered using Alameda Creek for a public waterworks, but city leaders were discouraged when they discovered that a dispute over title to the creek would interfere with their plans.

short in local creeks. Despite these promising circumstances, several barriers still impeded public development. Many voters remained unconvinced that public waterworks would save them money. The East Bay cities were so eager to protect home rule that they could not cooperate effectively. Finally, the Bay Cities Water Company did not have uncontested rights to Alameda Creek.

In 1905, East Bay residents had ample reason to complain about private water services. The previous year, engineers had reported that the "pressing need of additional water supplies in the East Bay cities is a matter of common knowledge."[81] A subsequent study blamed small water mains and scarce supplies for hindering industrial growth in the region. By the 1910s, many large water consumers, including hotels and factories, were going to extraordinary lengths to augment their supplies. The California and Hawaii Sugar refinery installed its own pipes and machinery to pump water from the Sacramento Delta, and, using barges and barrels, hauled additional supplies across the Bay from Marin County. The refinery spent $31,000 monthly on these complex arrangements. In addition to its own wells, Union Oil Company depended on ballast water carried by tankers from Seattle and Portland.[82]

While industry focused on the economic costs of inadequate water services, domestic users had their own water problems. In an unsigned letter, a Berkeley resident complained to the city council that water quality was unacceptable and that his water pressure was so low that "it does not throw the water over five feet from a garden hose when turned on to the full capacity."[83] J. B. Spears, janitor in Berkeley's city hall, protested that on at least two occasions "there was not sufficient water in the City Hall building to enable me to do my work of cleaning the building."[84] Such flaws of the private water service made public ownership seem all the more attractive to many East Bay residents.

The Bay Cities Water Company, a private firm, portrayed its water rights in Alameda Creek as the solution to Oakland's water problems. Because People's Water Company, Contra Costa Water Company's successor in the region, owned all other sizable streams in the East Bay, Alameda Creek was Oakland's last opportunity to build a municipal system with a local reservoir. Unfortunately, the Bay Cities proposal was deeply flawed. Even John L. Davie, a consistent advocate of municipal public works and Oakland's most prominent opponent of utility monopolies, judged the creek too small and polluted to meet Oakland's needs.[85] If he was correct, Oakland would have to purchase water and possibly also aqueducts and water mains from the People's Water Company.[86] The Bay Cities' water rights alone would have cost Oakland nearly six million dollars.[87] Buying additional water and a distribution system seemed prohibitively expensive.

Even more troubling, the Bay Cities Water Company did not hold undisputed rights to Alameda Creek. San Francisco's Spring Valley Water Company also claimed this source. In fact, the Oakland deal may have offered Bay Cities investors their last opportunity to profit from disputed water rights.[88] An effort by Oakland to develop Alameda Creek would probably have led to litigation with the Spring Valley Water Company that Oakland residents feared could cost their city dearly.[89] They also knew that the city could not expect to win easily. Several years earlier, the city of Oakland had faced the Contra Costa Water Company in court over the cost of water service. When the court decided in favor of the water company, the pro-utility *Oakland Tribune* accused the Oakland city government of wasting its resources on a fruitless suit. The *Tribune* maintained that the lawsuit had interfered with municipal services, and prevented Contra Costa Water from improving and extending its system.[90] Thus, according to the paper, municipal interference, not the private water companies, kept adequate water service from Oakland.

The sort of dispute over water rights that marred the Bay Cities proposal was quite common in California. By 1917, acrimony over the distribution of resources would prompt the state legislature to establish a state water commission with the authority to adjudicate water rights claims.[91] In Massachusetts, the conflicting demands of cities and industry also forced state officials to allocate water resources.[92] The main difference between Massachusetts and California in this regard grew from the relative place of private development and, in particular, the fact that California cities had weaker claims on water resources than their Massachusetts counterparts.

Despite the many flaws in the Bay Cities proposal, local leaders could not easily dismiss this opportunity for a municipal waterworks. Mayor Warren Olney, for example, saw the water issue as so important that he refused to support city hall construction, school improvements, fire protection, or parks until the water bonds passed. Business leaders felt that the mayor's inattention to these other public investments was shortsighted.[93] By pitting water against all other public works, Olney miscalculated. If he intended to bolster voters' resolve to pass the water bonds, he succeeded only in fragmenting support for municipal spending by appearing to hold the city's improvements hostage to water. Moreover, because of state-imposed debt and spending limits, Oakland could not pursue the Bay Cities plan and at the same time undertake other projects.[94] In 1905, these limits severely restricted municipal enterprise and hindered Oakland's efforts to satisfy diverse demands for public services.

As support for the Bay Cities proposal eroded, People's Water Company further complicated public water debates by offering to sell its reservoirs and distribution network to the city.[95] This option actually had greater merit than the Alameda Creek plan. Purchasing a complete system would have allowed the city to begin water deliveries immediately. The opportunity to expand an existing network, rather than start an entirely new one, particularly appealed to those who believed Alameda Creek was too small. Unfortunately, People's Water was only offering voters a red herring. Despite Oakland's interest, the company refused to name its price, so the purchase could not proceed.[96]

In December 1905, voters narrowly defeated the Bay Cities plan.[97] As defeat of the Bay Cities proposal demonstrates, the combination of state-imposed debt limits and municipal commitment to low taxes hobbled municipal efforts to expand services. But more importantly, this episode underscores the way scarce water resources influenced the history of public services in California. While water shortages led some voters to endorse municipal ownership, they also made water service profitable. The potential for profit, in turn, encouraged private utilities to monopolize key resources. After rejecting Alameda Creek, the city had no alternatives for public development other than purchasing a water company outright, or building a lengthy and expensive aqueduct to a distant river. Both of these solutions seemed impossibly expensive for Oakland to undertake alone. The combination of cost and resource scarcity pushed Oakland to consider interlocal cooperation.

Hetch Hetchy and San Pablo Dam

San Francisco, in the early 1900s, faced many of the same water problems as its neighbors across the Bay. The Spring Valley Water Company, which owned all local supplies, charged high rates but provided inadequate service. Given this monopoly, by 1908 San Francisco had abandoned plans for public waterworks drawing on local supplies; instead the city looked to the Sierra Nevada for a public water source. Municipal development of distant rivers was not a particularly original idea. Many cities, including Los Angeles, New York, and Boston, sent aqueducts out like tentacles to grasp distant waters. The East Bay's own water companies mentioned this option from time to time as well. But developing a Sierra supply lay beyond the means of any single East Bay city, at least without a dramatic increase in public spending. Participation in another city's Sierra water development, however, promised a less costly solution. Oakland welcomed San Francisco's plans to build a reservoir on the Tuolumne River in Yosemite National Park's Hetch Hetchy Valley. Hetch Hetchy offered the East Bay copious water independent of utility company control. With water from Hetch Hetchy, Oakland could build a municipal waterworks and free itself of shortages and water contamination.

The Army Corps of Engineers had investigated the water supply potential of the Tuolumne River between 1899 and 1900, and its report had attracted San Francisco's attention.[98] Not long afterward, Mayor James D. Phelan filed applications for rights to unappropriated storm runoff in the Tuolumne in order to preserve San Francisco's access to the river. In 1908, Secretary of the Interior James R. Garfield granted San Francisco a permit to dam the Tuolumne and create a reservoir in Hetch Hetchy. To this end, San Francisco had purchased nearly two million dollars worth of private property along the Tuolumne before the Sierra Club and Secretary of the Interior Richard A. Ballinger, Garfield's successor, mounted their opposition to the violation of a national park. Despite impassioned appeals to preserve the sanctity of national parks and the unique glory of Hetch Hetchy, Congress passed the Raker Act in 1913, authorizing San Francisco's use of the Tuolumne for a municipal water supply.[99]

Even before passage of the Raker Act, San Francisco had initiated discussions of interlocal cooperation on the Hetch Hetchy project. San Francisco's engineer, M. M. O'Shaughnessey, had courted East Bay support on the theory that the more people who benefited from the project, the more likely Congress would be to approve it. Initially, the East Bay did support Hetch Hetchy. In 1911, the Oakland city council declared its support for and desire to participate in the project.[100] The idea of cooperating with San Francisco appealed to Oakland's mayor, Frank K. Mott, and Berkeley's mayors, Beverly L. Hodgehead and J. Stitt Wilson, because the interlocal network would eliminate the need for multiple projects. A few decades earlier, a similar quest to improve services and eliminate duplication of effort had inspired Boston area communities to attempt similar partnerships.

Despite the initial appeal of transbay cooperation, East Bay enthusiasm quickly waned as Hetch Hetchy became identified with Bay Area metropolitan consolidation schemes. San Francisco boosters initially broached the idea of metropolitan consolidation in the aftermath of the 1906 earthquake and fire. By 1907, the San Francisco Chamber of Commerce had taken over the campaign, modeling their "Greater San Francisco" after the five boroughs of New York. The new metropolis was to include Colma, South San Francisco, and East Bay and Marin County cities.[101] In Oakland, consequently, Hetch Hetchy was widely condemned as an effort to reduce local autonomy.

Oakland's business community, in particular, strongly opposed Hetch Hetchy.[102] The thought that the East Bay's economic fate might one day lie in San Francisco's hands was anathema to ambitious business leaders who had long resented San Francisco's preeminence in northern California. For East Bay boosters, jealousy of San Francisco's relative importance was not merely a matter of civic pride, for image, investment, and industrial development went hand in hand. Significantly, East Bay opposition to Hetch Hetchy did not reflect sympathy with John Muir's campaign to preserve the valley. Indeed, those groups and individuals in the East Bay who most firmly rejected cooperation with San Francisco never attempted to prevent that city from building its reservoir. So, despite consistent support for the Hetch Hetchy project itself, vocal opposition to interlocal cooperation with San Francisco killed East Bay participation before city councils could even draw up formal proposals to vote on or put before the public.

Having rejected a transbay partnership, East Bay mayors now began to pursue their own regional water plans and to attempt greater control over private utilities. However, successive water company mergers greatly hindered their efforts by consolidating private ownership of water resources and creating a de facto regional system. Moreover, the mergers interfered with municipal efforts to regulate local water service because East Bay cities had little authority over activities outside their borders. Territorial constraints on municipal authority were once again impeding the control and development of public services.

Interlocal cooperation might have enabled the East Bay cities to develop, regulate, or purchase waterworks effectively, but San Francisco's shadow lingered, making cooperation even within the East Bay seem unacceptably risky. By 1910, many East Bay residents suspected that all cooperative projects concealed plans to consolidate their communities with San Francisco. In the 1870s and 1880s, Medford had reacted to Boston's Mystic Valley Sewer with similar suspicions. In Medford's eyes, all of Boston's actions seemed calculated to give Boston control over businesses and municipal services in the valley. Like Medford, Oakland and the other East Bay cities were suspicious of their larger neighbors and reacted by defending local independence vigilantly, even when that prevented them from building necessary public services.

In the years after it rejected Hetch Hetchy, the East Bay found its water problems growing steadily worse. Between 1908 and 1916, People's

Water Company repeatedly failed to increase supplies. Eventually, it did complete a new dam and reservoir on San Pablo Creek which, coupled with strict water conservation and record rainfall, temporarily reduced the sense of emergency.[103] When drought returned in 1918, war industries had inflated East Bay water use by 3,000,000 gallons of water a day. The water company's failure to increase supplies now became obvious, and pressure for public waterworks mounted once more.

Although they stopped short of blaming the drought on the water company, critics found fault with nearly all water company policies. High water rates were a particular irritant. Because the California Railroad Commission based utility rates on the total value of a company's property, a furor arose in 1917 when the East Bay cities accused People's Water Company of overestimating the value of its property in order to charge customers more.[104] In this light, the San Pablo Dam project became a "ruse of the octopus to secure a higher water rate" and evidence of "the rapacious greed of a soulless water trust."[105] Water bills were a constant reminder of the power of service monopolies, and of municipal government's lack of direct control over public services.

High water rates and poor service sparked interest in specific public water proposals; but anti-utility sentiment pervaded all the East Bay's battles with the private companies. The 1917 allusion to People's Water Company as the "octopus" demonstrated the extent to which Californians equated all utilities with the much reviled railroads. This same perception reappeared some years later, when the *Oakland Press* asserted that an unpopular water merger took place "under the parental wing and guiding influence of the Southern Pacific."[106] Although no direct ties had existed between the East Bay's water and railroad companies for many decades, opponents of the water companies continued to associate the two.

By the 1880s, antirailroad sentiment was a well-established characteristic of both political parties in California.[107] In their 1902 platform, California Democrats declared, "we denounce Private monopoly in every form," and argued that trusts would, "if not effectually checked, . . . prove subversive of the government."[108] The Progressive weekly *Pacific Municipalities* frequently echoed these sentiments, associating private utilities with tyranny, unrestrained corruption and an end to municipal democracy.[109] Eventually political candidates adopted the municipal water cause as a campaign strategy much as Hiram Johnson focused on railroad corruption and regulation during his 1910 gubernatorial bid.[110] Progressives came to dominate California reform politics in the 1910s largely by adopting this antirailroad stance.[111] As the Bay Area became a major stronghold of theirs, Progressives increasingly turned their attention to the political influence of urban utilities.[112] While reformers in the nation's eastern cities blamed immigrant-supported machine politics for corruption, Progressives in California targeted the railroad and private utilities as the sources of all political evil.

The Limits of Municipal Enterprise

For Boston and Oakland, the benefits of nineteenth-century municipal initiative did not last long. Improved waste disposal and access to new water supplies had allowed cities to overcome environmental limits to growth. Service improvements also spawned unanticipated service demands. As a result, when cities or private utilities expanded urban infrastructure, increased use of resources quickly pushed these systems to the point of collapse. Thus, the very success of municipal projects led to their downfall. Critics did not see this connection, however. Instead, they held up service problems as evidence of political corruption or municipal incompetence, and endeavored to isolate public works from traditional governance.[113] But before they abandoned municipal enterprise entirely, city leaders attempted to improve services within established institutions. Interlocal cooperation was among the most promising, and thus disappointing, of these experiments. When interlocal projects failed or were rejected, as occurred repeatedly in Boston and the East Bay, municipal oversight of services was further discredited. This, in turn, pushed Boston and Oakland ever closer to regionalism.

Boston had substantially more success than Oakland in establishing public authority over key natural resources. As a result, Boston built more effective and elaborate water and sewer projects, and exercised more political influence outside city limits. But in both cities, municipal authority over public works began with those services that everyone wanted but no one wanted to own, because services that generated revenue made good candidates for private development. Many services now commonly associated with public enterprise began as profitable businesses which governments took over when changing technology or business climates rendered them unprofitable. Waste disposal clearly demonstrates this transition from private to public responsibility. Until the mid-nineteenth century, private scavengers and privy-vault cleaners supported themselves collecting urban waste and selling it to suburban farms. Once running water fundamentally altered the nature of household wastes and suburban growth reduced the market for nightsoil, sewage disposal no longer supported the private waste collectors. At that point, urban governments assumed the task. By the 1860s, when Oakland built its first sewers, few questioned the necessity for public sewers. American cities had asserted their authority over waste disposal, or more accurately, had had sanitation thrust upon them.

Because water companies could earn a profit, urban residents approached public waterworks with less enthusiasm than they did waste disposal. This hesitation appeared clearly in Oakland, where scarce resources and overall aridity helped private companies establish resource monopolies. Because their customers had so few alternatives, East Bay water companies maintained their position for many decades. Although private companies did exist in Boston, they never monopolized water supplies and thus could not mount an effective opposition to public waterworks. But in both cities, the utilities opposed municipal efforts to take over their businesses. They found allies among citizens who preferred private to public enterprise, or who appreciated the low taxes and small government possible with private services. The East Bay cities confronted their private water utilities at a time when small government was particularly popular. This fact, their limited opportunities to assert a greater public interest in municipal ownership, and the continued profitability of private water all constrained public water authority in the East Bay.

The expansion of public authority in Oakland was further constrained by widespread fear of concentrated power. In the East Bay, antipathy towards centralization created strong opposition to both public and private services. On the one hand, public waterworks were widely seen as the only way to eliminate the menace of utility monopolies, a menace that seemed particularly threatening when the utilities were suspected of manipulating city officials. On the other hand, municipal water itself could "be turned into a political machine" that could institutionalize rather than eliminate political corruption.[114] For their part, Bostonians feared that increased public spending would breed graft; this prompted city officials to isolate water and sewer administration from electoral politics.

Despite concerns over corruption, Boston had several crucial advantages over Oakland when it came to establishing public services. First, Massachusetts laws reserved large lakes, known as "Great Ponds," for public use. Therefore, Boston had no trouble claiming water supplies for municipal development. Second, the strong association between public health, clean water, and good sanitation defined many services as essential to the mission of municipal government. The associations among disease, immorality, and filth evolved during the mid-nineteenth century into a complex philosophy of human behavior and the physical environment, readily adopted by social reformers, health professionals, and politicians. Moral-environmental philosophy lent particular force to municipal sanitary measures by vesting water and sewer projects with purported power to eliminate social degradation. East Bay leaders limited their campaigns to a much narrower and contentious arena. They railed against the negative influence of monopolistic utilities, and their call to reform government and ensure economic growth held little of the profound, utopian promise of Boston's municipal sanitation campaigns. And while Oakland residents welcomed services and projects that promised to increase their community's prominence, independence, and commercial development, the idea of public ownership in Oakland never achieved the same cure-all status that moral-environmental projects had in Boston.

Despite their early promise, water and sewer systems in both Boston and Oakland quickly revealed profound weaknesses. By the 1910s, Oakland's local water supplies had proved inadequate. The private water company's acquisition of waterworks in neighboring communities impeded Oakland's regulatory efforts by placing much of the water system outside the administrative reach of city officials. Even so, consolidation of water supplies failed to prevent severe shortages. In Boston, extensive public sewerage and water supply improvements proved similarly disappointing. Poor waste disposal practices in the Mystic Valley and Charles River in particular limited the effectiveness of Boston's public works investments. Increased urban development in both the Boston and Oakland metropolitan areas overwhelmed independent municipal efforts to provide and regulate services. These developments demonstrated the limitations of municipal governments in addressing problems that originated outside city boundaries and left city administrations vulnerable to accusations of graft, inefficiency, and poor planning. Indeed, late-nineteenth and early-twentieth-century criticism of municipal administrations was closely associated with the decline of the first generation of municipal public works projects. This decline of municipal systems should not be interpreted as an indication of the inadequacies of the these networks, but rather as the political manifestation of the limitations of local solutions to environmental problems that extended far beyond municipal borders.

3 | Boston: Regionalism in the Gilded Age

By 1890, Boston had spent nearly a hundred years building water and sewer networks. Systems that had begun as relatively modest municipal projects had expanded to accommodate population growth, suburban annexation, and dramatic increases in water consumption. In some ways, these municipal enterprises had been quite successful. The city water department supplied residents with adequate, inexpensive water service. Sewers, particularly the Main Drain, kept sewage and storm water out of basements and off city streets. Given the improvements in domestic plumbing and rising expectations of public services that had taken place, this was quite an accomplishment. The fact that water and sewerage represented a mere fraction of Boston's municipal projects makes these accomplishments all the more impressive.

The average Bostonian living in the Back Bay or drawing water from the Mystic waterworks in the 1890s, however, was less impressed. Despite Boston's response to demands for water and sewerage systems, a number of highly visible and apparently chronic problems still plagued the city. While Boston did not face actual water shortages, ever-increasing water use prevented the city from abandoning its polluted sources. Suburban growth now threatened even the previously pure Lake Cochituate. Despite completion of the Main Drain in 1884, wastes from the northern suburbs still flowed past Boston on the Charles River. Odors from the South Bay and Roxbury Canal remained an ever-present threat. These problems were, of course, essentially the ones that the city had tried to solve with annexation and interlocal cooperation in the 1870s. The fact that sewage nuisances and water contamination not only persisted but increased in the 1880s and 1890s discredited municipal enterprise.

Recognizing that Boston residents were thoroughly frustrated with municipal efforts, state officials and political reformers seized the opportunity to find new, regional solutions. In the 1880s and 1890s, opponents of the Democratic Party, including the State Board of Health and a variety of Yankee, Mugwump reformers, created metropolitan sewerage and water agencies for the Boston area that gained the support of voters who might otherwise not have endorsed the political reforms necessary for centralized administration of public works.

Boston's metropolitan—or regional—water and sewerage were built and administered by special districts. Since special districts were not municipal governments, they were exempt from limits on the amount of debt a municipality could incur and the level of taxation it could impose—limits that had been enacted by state legislatures after a rash of mid-nineteenth-century railroad bond defaults. Public works advocates realized that creating a special district would enable them to increase public spending without exceeding caps on city budgets and taxation. These financial advantages increased with the advent of the revenue bond, which permitted a special district to borrow against its future income instead of assessed property values. Although Americans had used special districts to provide rural services such as schools, parks, and transportation, the idea of creating special districts to provide services to metropolitan—as opposed to municipal or rural—areas represented a significant evolution in American government.[1]

For Boston, regionalism made sense in a number of ways. Central coordination of water resources, whether for waste disposal or drinking supplies, meant reduced competition and pooled finances for expensive construction. To a certain extent, it had also eliminated conflicting resource uses. All of these accomplishments had been beyond the scope of municipal development. Regionalism also made sense in political terms. The planning and coordination inherent in these projects permitted the application of technology and professionalism on a nearly unprecedented level, to the satisfaction of those entranced by the power of science and the promise of professional expertise. Construction created jobs, which pleased some constituents. Institutional innovations appealed to political reformers who wanted to limit municipal power. Meanwhile, the fact that special districts neither had direct oversight over existing city institutions nor required cities to abandon or transform their traditional governmental structures met with approval from municipal officials. That regionalism satisfied so many different groups, and addressed so many of the reform priorities of the late-nineteenth century, accounts for its popularity and the relative ease with which Boston accepted the Metropolitan Sewerage Commission and Metropolitan Water Board.

The Metropolitan Sewerage System

In the 1880s and 1890s, Boston transferred a number of municipal responsibilities to regional agencies, beginning with sewers. The creation of the Metropolitan Sewerage System in 1889 represented a major transition in Boston's politics. Political reformers seized upon the opportunity provided by chronic drainage problems and contaminated reservoirs to discredit city government in general and municipal officials in particular. They capitalized on the appeal of improved, sewers—and, later, other services—to create new, centralized institutions that they would administer. Thus Republican Mugwumps were finally able to seize power from their Democratic rivals. The State Board of Health pursued regionalism aggressively, a stance that reflected its members' connections to political reform movements as well as its mission to eliminate the unsanitary conditions that bred disease. Where earlier attacks on municipal governance had failed, regionalism succeeded because its promise of improved services and offer of employment on public works appealed to voters and their municipal representatives as strongly as new institutions and efficient, "apolitical" administration appealed to the reformers.

Mystic River

The sewage crisis in the Mystic Valley, which was a major source of Boston's water and home to a number of independent-minded communities and many polluting industries, exemplifies the way conflicting water uses and interlocal rivalries drove Boston to adopt regionalism. As described in Chapter Two, political constraints prevented Boston from building a sewer to collect both domestic and industrial wastes. So, even with the Mystic Valley Sewer in place, domestic wastes and the effluent from several tanneries and at least one glue factory still polluted Boston's reservoir.[2] Moreover, despite legislation requiring the city to purify wastes from the sewer, water quality in the Mystic Lower Pond had not improved significantly. By 1881, the city council readily admitted that the Mystic Valley Sewer failed to protect water supplies. Boston Alderman Charles H. Hersey became so frustrated with the Mystic that he suggested abandoning the network entirely and building a new water supply.[3] Persistent problems in the Mystic made Boston's leaders and voters quite receptive to regional approaches to sanitation.

The State Board of Health was instrumental in identifying the problems in the Mystic Valley. Despite its limited enforcement powers, the Board tried to intercede in sanitary planning. At times, however, its oversight caused more problems than it solved. Citing the hazards of polluted drinking water, in 1873, the State Board of Health vetoed Medford's local sewerage plans, but did not offer Medford an alternative.[4] The town blamed this interference on Boston rather than on state officials. That their sacrifice might prevent disease in Boston, of course, was of little concern to Medford's leaders.

The State Board of Health placed Boston under similar restrictions. In 1874, it refused to allow Boston to build an outfall into the lower river.[5] In many ways, the lower, tidal section of the river would have made a better sewer outfall than Mystic Lower Pond. If the Board had permitted Boston to end its sewers below the dams, at least some of the waste that concentrated in the lake would have washed into the harbor. Possibly the Board issued its prohibition on river disposal to protect downstream communities from the kinds of concentrations of sewage that afflicted the Charles River. Ultimately, of course, wastes accumulating in the lower pond caused more obvious problems than they would have if dumped below the dam.

The State Board of Health's decrees and Boston's own efforts to protect its reservoir led eventually to the fouling of Mystic Lower Pond. In response, in March 1881 Governor John D. Long used his inaugural address to call for a state investigation of "a comprehensive system for draining the entire area . . . within . . . ten miles from the State House." He believed that contamination of the two rivers would only increase, and that "the foul condition of either stream will be an injury to both health and comfort in the towns near its mouth, however good their own drainage may be."[6] Governor Long made clear that independent municipal action could not eliminate sewage pollution and that he understood that the situation required state intervention. The General Court responded, easily passing a resolution for a thorough investigation of the Mystic and Charles rivers' drainage.[7]

Acting quickly on the General Court's resolution, Governor Long appointed a Metropolitan Drainage Commission to conduct the study. The five-member commission included both Chesbrough and Folsom who, of course, had already demonstrated their commitment to regional sewerage planning. The Drainage Commission's final report argued that only comprehensive sewerage for the entire Boston basin could preserve local water quality for industrial, recreational, and domestic uses. To this end, they recommended that the state build a network of interceptors and sewer mains to collect town and city wastes for the Boston metropolitan district.[8] Many other studies and reports would follow, but nearly all drew the same conclusion: Boston needed a metropolitan sewerage district with a regional network of drainage pipes.

By 1882, the Massachusetts State Board of Health, Lunacy and Charity, as the health agency was then called, had committed itself to regionalism.[9] The agency's annual report declared that "the remedy which will finally relieve the city of Boston from its . . . anxiety as to the protection of the Mystic water-supply . . . lies in a comprehensive system of drainage for the whole Mystic valley."[10] The Board of Health did not have the authority to require towns to cooperate. However, prohibitions against waste disposal in the river, such as those that prevented Medford from improving domestic sanitation, seemed calculated to force the Mystic Valley to adopt centralized sewerage in some form. By prohibiting river disposal, health officials compounded the valley's sewage crisis and exacerbated intergovernmental conflicts thus engendering the frustration that eventually pushed Boston, Medford, and other neighbors to embrace the Board of Health's vision.

Natick and Lake Cochituate

Mystic Lake was not the only water supply plagued with sewage contamination. By the time Governor Long called for drainage studies, industrial and residential growth in Natick had rendered Lake Cochituate increasingly unsavory. As in the Mystic Valley, incompatible water use lay at the heart of the matter. And, as in the Mystic, state prohibitions limited Natick's and Boston's options when they sought to reconcile water supply with waste disposal.

When Boston first built the Cochituate system, the lake's rural setting and isolation from concentrations of industry and housing assured the city that its water supply would long remain pure. Protection from possible contaminants was a particular priority in the 1840s, because no practical means existed to remove impurities from large quantities of drinking water. Eventually, the extensive watershed that made Cochituate so attractive as a water supply became its greatest liability. By 1872, Natick's private industrial and domestic sewers drained enough wastes into Lake Cochituate to cause concern.[11] Natick residents needed sewers and wanted running water, but those services would increase the wastes reaching Lake Cochituate many times over. Boston health officials despaired of halting Natick's plans, lamenting that they had no recourse but to "quietly observe the gradual deterioration of the water, until actual sickness and death of water takers make an injunction possible."[12] By then, of course, the reservoir would have been hopelessly contaminated.

In response to the problems brewing around Lake Cochituate, in 1875 the mayors of Natick and Boston together petitioned the General Court to preserve Cochituate's purity. The legislature responded, granting the Massachusetts Supreme Court the authority to prohibit the discharge of sewage or other contaminants into Lake Cochituate and its tributaries. The act also permitted Natick to divert Pegan Brook, the main sewage-bearing tributary, away from Cochituate so long as the town did not send its wastes into the Charles River.[13] Natick and Boston implemented this legislation by installing filters to purify water entering Cochituate from Natick, but this solution eventually proved insufficient. The same 1882 metropolitan drainage study that urged regionalism in the Mystic Valley cited sewage contamination in Lake Cochituate as yet another reason for centralized sanitation.[14] By 1885, Pegan Brook ran sluggishly and emitted "at times a bad odor, caused . . . by sewage which discharges into it."[15] Natick and Boston feared that the filters did not remove "liquid impurities" from the brook and would ultimately fail to protect the water supply.[16] The patchwork of well-intentioned sewage disposal rules applied only to new sewer projects. Therefore, they did little to reduce pollution from older sources, and in many cases forced communities to dispose of their wastes in clearly unsanitary ways. Ultimately, the state restrictions that prevented Natick and Boston from diverting wastes out of the Cochituate watershed had the same effect there as did the ones in the Mystic Valley.

There is less direct evidence than in the Mystic Valley case that state leaders intended their Cochituate rulings to force Boston and Natick to embrace coordinated sewerage. Nevertheless, that is more or less what happened. Eventually, Boston extended the Main Drain to Natick, diverting Natick's wastes away from the reservoir. Incorporating Natick sewerage into a system that crossed municipal lines solved water contamination problems that lesser interlocal and municipal initiatives could not. Therefore, this action provided further evidence of the value of centralized, multiple-city service administration as a means to overcome metropolitan environmental crises.

Charles River

Contamination dogging the Charles River received even more attention than did similar problems in the Mystic River and Lake Cochituate. Although the Charles did not supply drinking water to Boston, there was continued concern that foul odors, not just contaminated drinking water, threatened public health. Moreover, the Charles flowed past the neighborhoods of some of Boston's most influential citizens. For reasons of familiarity and politics, then, Charles River pollution dominated discussions of Boston's sanitation.

The focus on the Charles began early. As seen previously, the Main Drain grew largely from efforts to clean up the Charles riverbank. It diverted most of Boston's sewerage into deep water and eliminated the low-gradient drains, tide-locked sewers, and waterfront outfalls that had made the Back Bay so noisome. But, of course, not all of the sewage in the Charles originated in Boston. Cambridge, Somerville, and the industrial communities of Waltham and Watertown poured their effluent into the stream, overwhelming any water quality improvements from Boston's project.[17] Frustrated Back Bay residents were especially active, sending petitions, complaints, and reports to the city council and legislature almost unceasingly from 1877 to the end of the century. Their continued dissatisfaction with conditions in the Charles kept sewerage on the public agenda and highlighted the need to find remedies that extended beyond municipal borders.

The State Board of Health advocated regionalism to remedy contamination in the Charles, as it had elsewhere. In 1878, as Boston began building the Main Drain, several Board studies concluded that only coordinated drainage would improve Charles River conditions.[18] The Board was explicit that mere interlocal cooperation would not suffice, proclaiming it "nearly impossible for different municipalities, whose interests are in reality . . . conflicting, to act in harmony with each other in matters in which both are concerned." It went on to assert that "if the whole question of sewerage of the suburbs of Boston were controlled by one authority," sanitation problems in the Charles might finally be solved.[19] Pointing to Cambridge's woeful history of failed interlocal sewerage construction, the State Board of Health concluded that further municipal sanitation efforts in Cambridge would prove no more successful than those in Boston.

In April 1882, the State Board of Health notified the Boston City Council that it had received a number of petitions regarding the "offensive

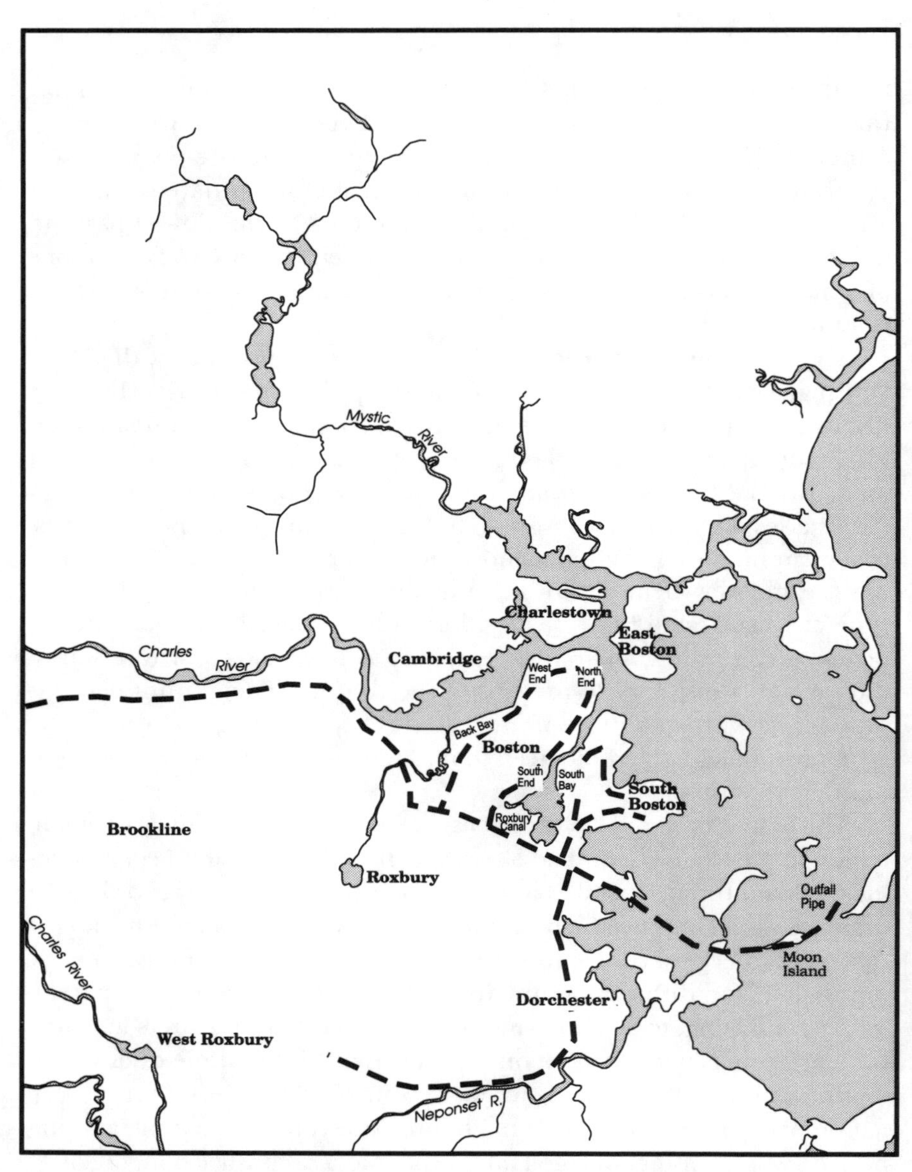

Boston's metropolitan drainage. Approved in 1889, the metropolitan sewerage system expanded the municipal main drain and laid the groundwork for comprehensive sewerage for the whole Boston basin.

condition of the flats" in the Charles. Residents complained that "sewer deposits . . . [had] accumulated to such an extent as to occasion serious annoyance to the inhabitants in the vicinity."[20] Health officials blamed the growing problem on new starch and woolen factories disposing of "considerable quantities of putrescible organic matter" in the river. Now problems were no longer confined to the sections of the river nearest Boston. The Board found that a "foul odor pervaded the entire neighborhood of the river between Watertown and Newton."[21]

In 1884 and 1885, just as Boston completed the Main Drain, state-appointed committees continued to research sewage nuisances and to recommend regional, coordinated drainage. The persistence of nuisance conditions in the Charles after completion of the Main Drain reinforced conclusions reached about the Mystic and Cochituate watersheds, confirming the need for a new institution to manage Boston area drainage. As one sewerage commission reasoned some years later, "Any sewers constructed for the relief of one . . . would necessarily traverse one or more districts under another jurisdiction; and . . . no authority, other than that of the Commonwealth, was sufficiently comprehensive to embrace the entire district . . . ."[22] As the 1880s wore on, state and local officials were no longer content to suggest that Boston coordinate its sewerage with its neighbors. Particularly in discussions of Charles River contamination, proponents of coordinated sanitation explicitly advocated the creation of a new political agency with authority over all sewerage throughout the Boston area. This recommendation marked an important departure from earlier reports. Creating a new political entity was to have important implications for municipal politics and intergovernmental relations in the late 1880s.

Politics of Regional Sewers

In June 1889, the Massachusetts General Court passed the "Act to Provide for the Building, Maintenance and Operation of a System of Sewage Disposal for the Mystic and Charles River Valleys," and thus created the Metropolitan Sewerage Commission.[23] The measure passed with remarkably little controversy. In fact, the legislators who demurred did so on the grounds that the system was too small rather than too ambitious.[24] For their part, most municipalities welcomed the new institution. Having exhausted local remedies, they embraced regionalism as a much-needed solution to exasperating problems, and as one that did not appear to threaten home rule because it left most municipal responsibilities and institutions intact. A broad coalition of state and local leaders endorsed regionalism. Citizens, too, welcomed the new institution, expecting a wide range of social, political, and economic benefits from new sewerage construction.

The Metropolitan Sewerage Commission assumed responsibility for Boston's Main Drain and began new construction almost immediately. By 1895, the commission had completed the area's first truly regional drainage system. This network of intercepting sewers, called the North Metropolitan Main Drainage, diverted wastes from the Mystic River and the north bank of the Charles into Boston Harbor. For this and subsequent projects, regional sewer planners adhered closely to the recommendations that Chesbrough, Lane, and Folsom had made in their 1876 Boston sewerage report. In 1899, the General Court approved the South Metropolitan Main Drain to improve conditions in Boston's third and most industrial river, the Neponset.[25] A center for manufacturing since the seventeenth century, the stream had suffered from many of the same kinds of problems that afflicted the Mystic and Charles Rivers.[26] By 1905, almost all of the Boston Harbor watershed was linked by intercepting sewers administered by the Metropolitan Sewerage Commission. With the founding of the Commission in 1889, not only had Bostonians fully accepted regionalism, but they had adopted a model for comprehensive planning and state administration that they soon applied to other services, including parks in 1893 and water supply in 1895.

The creation of the Metropolitan Sewerage Commission did more than provide Boston leaders with a new administrative gimmick. Urban regionalism transferred real power from elected municipal officials to their political rivals. Moreover, this transfer of power took place with the blessing of the very groups who lost the most influence as a result of regionalism. The fact that the Metropolitan Sewerage Commission sparked little controversy is all the more significant and surprising given the ethnic and class lines that divided Boston in the late nineteenth century.

By the 1880s, Boston's ethnic groups strongly influenced municipal and party politics at least at the ward level. African Americans wielded significant power in several South End wards. Committed to the Republican Party, they endorsed candidates on their antebellum abolitionist record, and by the 1870s had elected black representatives to both the city council and the state legislature.[27] The Irish, plentiful throughout much of the city, provided the Democratic Party with a large enough

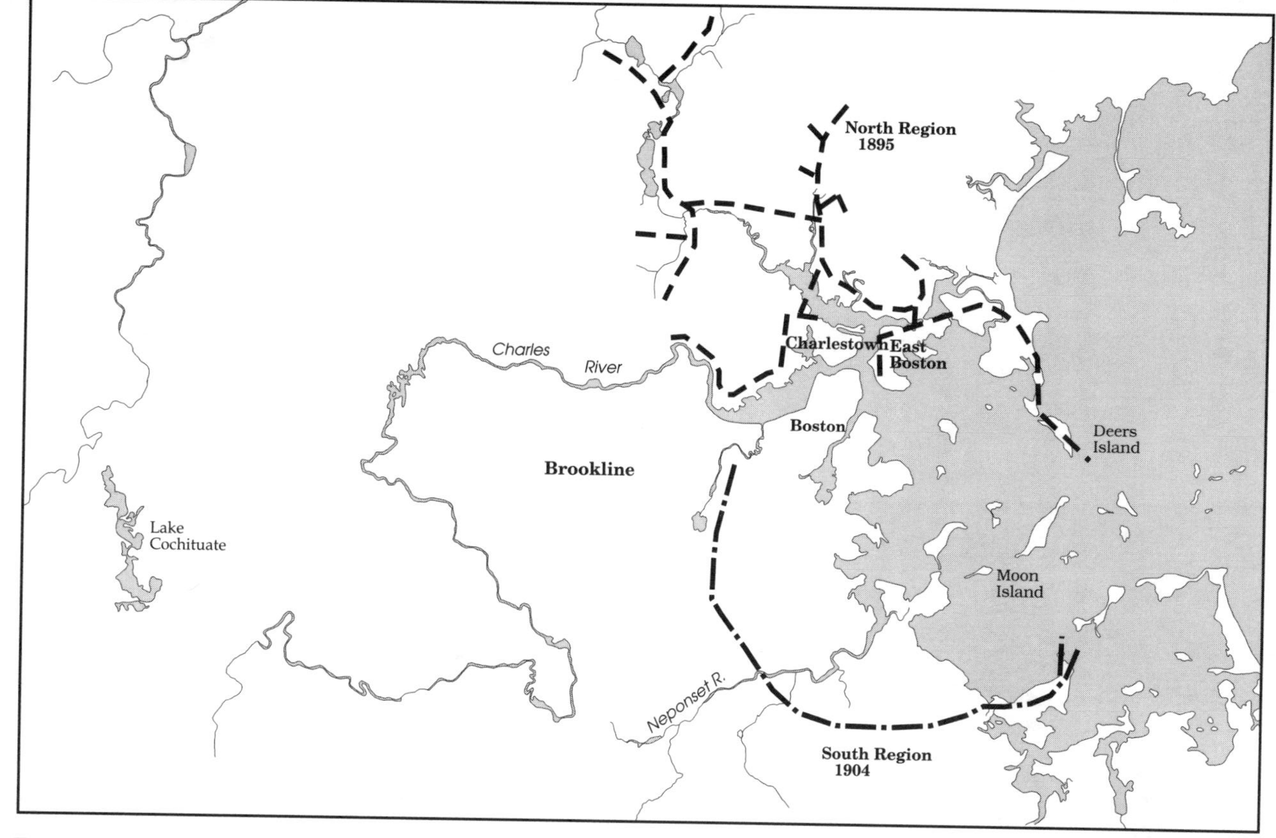

Expansion of the metropolitan drainage system. In 1895, the Metropolitan Drainage Commission began building the North Region sewers to eliminate nuisances in the Mystic and Charles rivers. A few years later they established the South Region to address pollution in the Neponset valley.

Moving shielding for the High Level Sewer, a southerly addition to the Metropolitan Sewage System. Boston, 1901. Courtesy of the Massachusetts Archives, Boston, Massachusetts.

Laying sewer lines in a Malden neighborhood as part of the North Region of the Metropolitan Sewage System, 1906. Courtesy of the Massachusetts Archives, Boston, Massachusetts.

constituency to dominate City Hall. They were quite aware of their importance to the party, as the following editorial comment in the city's Irish newspaper, *The Boston Pilot,* demonstrates: "The Democratic party of Massachusetts—what is it? Take away the Irish element, *where* is it? No where."[28] Irish voters at this time usually elected prominent Yankee Democrats to office, but these leaders did not always share Irish priorities. In an impatient moment, editors at *The Pilot* complained, "It is all well enough so long as we do the dirty work for the Democracy, but the moment we ask for anything for one of our own race, we are met with the cry, 'He's Irish' and . . . 'No Irish need apply.'"[29] Further conflicts between Yankee and Irish Democrats eventually emerged because representatives from the Irish and working-class Democratic strongholds felt excluded both from party leadership and from the business of selecting candidates.[30]

The divisions within the parties paled in comparison to those that separated the Republicans and the Democrats. The Democratic Party secured the loyalty of its Irish membership with an efficient political machine. Employment on public works and promises of improved services, all essential features of the Metropolitan Sewerage Commission, were integral to the success of the Democratic organization.

In Republican eyes, the Democrat's electoral strategies bred corruption. They dismissed Democratic spending as wasteful, driven by private interest and "semi-communistic sentiment" rather than legitimate public needs.[31] But at the heart of Republican attacks on the Democrats and on Boston's municipal government lay an assumption that Democratic voters were themselves morally corrupt, and therefore incapable of electing any but corrupt leaders. This insinuation was partly an outgrowth of moral-environmentalism in which the moral corruption associated with filth and disorder had explicitly political implications. But reform was also motivated by nativism and anti-Catholic sentiment. For example, one Republican newspaper cautioned that "[t]he Roman Catholic Church . . . nourishes a one-man power not favored under our institutions" and implied that Catholics might not be able to adjust to the principles of a democratic system of government.[32]

In the 1880s and 1890s, the Republicans responded to the corruption they saw within the Democrat-dominated municipal government and to their own loss of political prominence with proposals to reduce municipal autonomy. Regionalism was merely one manifestation of this strategy. *The Boston Globe* stated this goal baldly when it asked readers if "it may not be well for us to take some steps toward securing a more equitable adjustment of administrative control" than that provided by universal suffrage.[33] The Republican *Boston Journal* was even more explicit, suggesting that special commissions take over all administrative duties and, thereby, transform the city council into a purely legislative body.[34] This would have severely restricted voter influence in local government. In 1895, Republicans promoted annexation of Chelsea, Revere, and Winthrop as means to counter Democratic power in Boston, and transform the city into a "Republican stronghold."[35] All of these Republican ideas revealed frustration with popular sovereignty and a defensive effort to maintain elite power.

Voters usually rejected Republican reform proposals precisely because they did interfere with municipal autonomy or the structure of city government. Democrats were particularly critical of special legislation; they felt that Republican legislatures "singled out" Boston "year after year . . . as a city that was terribly in need of reform . . . always under the guise of good government, but manifestly in the interests of the Republican party."[36] The Democratic *Boston Post* complained as early as 1881 that increased centralization threatened local autonomy throughout the nation, asserting that "the rightful authority of the States has been seriously interfered with by successive national administrations that have perverted the theory upon which our peculiar system of government was founded, and the States in turn . . . have had their revenge by taking the large cities under their peculiar care." Moreover, Boston had "felt the injustice of being subject to the ignorance of our needs, and, frequently, prejudice against our institutions that prevails among a large portion of the Massachusetts legislatures . . . [and] is troubled with more outside pressure than is good for her."[37] Newspaper publisher and influential Irish Democratic leader Patrick Maguire denounced as "venerable hayseed legislators from Podunk and other rural sections" and "Boston hypocrites and frauds" the legislators who sought to increase state supervision of Boston municipal administration.[38]

Significantly, neither Maguire nor other protectors of Boston's autonomy saw regional sewers in quite the same light as they did these other reform projects. The expansion of Boston's water and sewerage systems in the decades leading up to regionalism set ample precedents not only for bipartisan endorsement of public works, but also for faith in semiautonomous public service boards and commissions. Moreover, the suburban communities that consistently resisted Boston's expanding influence, including Medford, also accepted centralized sewerage with little resistance. Even the most vocal protectors of local political institutions appeared more than willing to surrender sanitation, and later water supply, to state or regional control in return for the benefits they expected to reap from the new services. Lingering notions of the moral-environmental hazards of unsanitary urban conditions obviously played an important role here. Municipal leaders welcomed the opportunity to provide jobs for working-class constituents and public investment to boost the commercial enterprises of the entrepreneurial class.

As Democrats and Republicans fought for political influence in Boston, state officials—particularly the State Board of Health—tried to tip the scales in favor of regionalism in other venues. The State Board of Health's efforts to reduce sewage pollution in Massachusetts rivers increased pressure for regionalism by forcing communities to find deep water outlets for their wastes. Board studies closely linking water contamination with inadequate sewerage gave many the impression that positive improvement in public health would elude municipalities as long as insisted upon single-city construction.

Despite this emphasis on regionalism and although the Board did become very much embroiled in partisan politics in the 1880s, the Massachusetts State Board of Health had a mixed record on municipal autonomy.[39] It did not restrict its suggestions for regionalism to urban areas dominated by Democratic "machines." Recommendations for regional systems included Middlesex County communities, such as Newton, Medford, Somerville, and Cambridge, which consistently elected Republicans to the General Court.[40] In fact, the Board advocated similar policies for communities of all sizes and political leanings. It may be argued that the Board could not wrest Boston's public works from voter control without including these towns, but the dimensions of the regional proposals—extending as they did so far beyond the area strictly necessary to justify state management of Boston's public works—contradict that notion.

The story of the Metropolitan Sewerage Commission demonstrates that the appeal of regionalism crossed the political divisions of ethnicity, class, and political party allegiance. The State Board of Health deserves much credit for both the urgency that carried regional sewers past partisan divisions and the popular conception of sewerage as a regional problem. But regionalism succeeded not so much by erasing partisan differences as by offering both Republicans and Democrats something that appealed to their constituents. Republicans and political reformers recognized that the Metropolitan Sewerage Commission reduced voter influence over public works and enshrined the principles of scientific management. But because regionalism combined these reforms with construction jobs and avoided direct interference in the institutions of municipal governance, regionalism also appealed to Democrats and others who remained dedicated to home rule. The metropolitan sewer district stirred little controversy because it so effectively addressed so many political concerns. As water quality in greater Boston continued to decline, the political lessons learned from sewerage were quickly applied to water supply.

Metropolitan Water Board

By 1895, six years after the Metropolitan Sewerage Commission was established, many in the Boston area realized that while regional sewerage solved many problems, it could not completely protect their drinking water from domestic or industrial drainage. Meanwhile, with competition for resources thwarting municipal efforts to develop new supplies, the best solution appeared to be regional water development of distant, rural water sources. The pursuit of a regional waterworks would bring rural distrust of cities into sharp relief, pit urban against rural communities, and project onto a larger stage the antiurban critique that had driven regionalism all along.

Urban growth was responsible for many of the water problems confronting Boston and its neighbors. As a result of industrial expansion and population growth, water use had increased steadily throughout the Boston area. In 1884, Boston's total water consumption had averaged sixty-five gallons a day per person; in every subsequent year, daily water use increased by two and a half gallons per person. By 1894, most Boston area towns used six to eight million gallons a day more than their reservoirs and wells could reliably supply.[41] Manufacturing accounted for a considerable portion of this consumption; the State Board of Health estimated that mills in Hyde Park, Mattapan, and East Dedham alone used twenty million gallons of water a day. According to reports issued in 1896, eight-five percent of the metropolitan population would need additional supplies by 1900, and ninety-seven percent would face severe shortages by 1920. Only Newton, Waltham, Cambridge, and Brookline could expect their systems to suffice beyond 1920.[42]

Water contamination compounded the problem of water shortages. Growing demand prevented many communities from abandoning low quality sources, while sewage pollution sullied many rivers and lakes that might otherwise have been used to increase water supplies. For example, Boston was forced to continue to use Mystic Lake long after the Boston Water Board and State Board of Health agreed that the reservoir should be condemned. Boston's need was so great and its opportunities to develop new reservoirs so rare by the 1890s, that the city council seriously discussed buying and closing the Mystic Valley tanneries. Meanwhile, Somerville, Everett, and Chelsea built filtration plants to purify the nearly undrinkable Mystic water.[43] Cambridge residents faced similar problems. In their case, there was a direct relationship between overdraft and increased contamination in Fresh Pond, Cambridge's reservoir. Many sewers in Cambridge and Somerville emptied into Alewife Brook. Until the late 1870s, the stream flowed north from Fresh Pond toward the Mystic River. As water use in Cambridge increased, water levels in Fresh Pond sometimes dropped low enough to reverse the flow of the brook, which then carried sewage back up into the reservoir.[44]

Most communities in the Boston area responded to the twin threats of water shortages and contamination by searching out new water sources. When they attempted to solve their water supply problems, however, city leaders throughout the Boston area confronted the familiar combination of incompatible resource uses, interlocal rivalry, fear of disease, and continued demand for improved services that had spawned the Metropolitan Sewerage Commission in 1889. But, despite the familiar scenario, supporters had to fight much harder for regional water than they had for the sewers. Some communities balked when they realized that regionalism would require them to abandon revenue-generating waterworks. Others felt that the taxation scheme penalized them or benefited their neighbors. Furthermore, as Boston communities reached beyond their borders into central Massachusetts in search of new water supplies, they aroused opposition in rural communities which compounded their difficulties.

Boston, Cambridge, and the Shawsheen River

As was the case with regional sewerage, the creation of the Metropolitan Water Board followed several decades of municipal efforts to improve water supply. Boston and its neighbors had pursued waterworks as aggressively as they could. From the 1870s on, their efforts had increasingly brought them into conflict with one another. The last major battle for municipal water supplies in the Boston area took place in 1882; Boston, Cambridge, and Andover entered a brief but intense dispute over the Shawsheen River. During this episode, cities not only competed for water with one another, but also with powerful industrial interests. The dispute over the Shawsheen set the stage for regionalism by proving to Boston area communities that they could no longer rely on local rivers.

The Shawsheen rises between Lexington and Bedford, west and north of Boston. From there, it flows twenty miles north, to join the Merrimack River just below the old mill town of Lawrence.[45] Cambridge petitioned the General Court for permission to use the Shawsheen first. Shortages and contamination in Fresh Pond seemed to require it.

A private water company had begun selling water from Fresh Pond in 1856. In 1867, Cambridge purchased the pond and private waterworks; the development of Fresh Pond as a municipal water system roughly coincided with Charlestown's investment in Mystic Lake. In 1875, Cambridge sought legislative permission to increase the Fresh Pond supply by connecting the main reservoir to Spy Pond in Arlington, Little Pond in Belmont, Alewife Brook, and its tributary Wellington Brook. In 1878, before the city took any action, water quality in Fresh Pond declined suddenly. City officials traced the contamination to a slaughterhouse operating on the shores of the lake in Belmont. Convinced that they needed expert help, they sought advice from Ellis S. Chesbrough before expanding the Fresh Pond system. Chesbrough judged water quality in all the ponds and brooks that Cambridge hoped to use to be too poor for domestic use. Cambridge's own sewers contributed significantly to these problems. To preserve water quality, Cambridge discovered it had to stop using water from over half of the Fresh Pond watershed. At the same time, the city needed an additional 750,000 gallons of water per day just to meet existing demands. In 1880 and 1882, dry weather forced Cambridge to seek supplemental supplies from Boston, but Boston rebuffed the request, claiming it had no excess water to sell.[46]

In 1881, after Boston refused its request for water, Cambridge petitioned the General Court for permission to divert water from the upper reaches of the Shawsheen River, through the old Middlesex Canal and into Fresh Pond.[47] The Shawsheen promised pure water in large enough quantities to meet Cambridge's existing and future needs. If the Cambridge city council anticipated a rapid decision in their favor, however, they were disappointed. Textile and machine tool mills in Andover and

North Andover, the two communities located closest to the Shawsheen's banks, relied heavily on the river for water, power, and waste disposal. They were not willing to give up their claims to the river.

Before Cambridge sent its petition to the General Court, Boston had paid scant attention to the Shawsheen River. In the 1860s, several years before it built the Sudbury River reservoirs, Boston had rejected proposals to supplement the Cochituate Water Works with water from the Shawsheen. The city had also rejected the Charles and the Mystic as possible water sources. Immediately following Cambridge's 1881 application, however, the Boston city council petitioned to use the river.[48] On this occasion, city councilor Malcolm S. Greenough urged his colleagues not to "stand by and see any other corporation come in and take the water of the Shawshine [sic] without making an effort to obtain it."[49] The idea that the Shawsheen could replace the still polluted Mystic was particularly appealing so soon after the Mystic Lower Pond incident.[50] It was clear that Boston had legitimate interest in the river, but as Greenough's statement makes clear, its petition was also an aggressive attempt to protect its control over local water resources.

Boston's petition transformed a two-way competition between the Shawsheen towns (backed by the powerful mill owners) and Cambridge into a more complicated battle. Both Cambridge and Boston requested allocations of Shawsheen water to meet urban domestic and industrial needs, but they disagreed on who should control or develop the river. Cambridge representatives argued during 1882 hearings on the Shawsheen proposals that an independent water supply was essential to their city's autonomy. They firmly rejected the idea of cooperating with or purchasing water from Boston, as either of these arrangements would place Cambridge under Boston's control.

Boston, on the other hand, proposed symbiotic development of the Shawsheen. But the city's motives were not entirely clear. Greenough's comments suggest that Boston officials may have proposed the joint project because they knew they had little chance of getting a municipal project through the General Court. In this case, the symbiotic project represented an effort to secure water rights that the city might otherwise lose. Despite the savings that they expected from joint development, Boston officials rarely emphasized the economic advantages of collaboration. Rather, they insisted that Boston needed the water to replace Mystic Lake and that the General Court should not prevent Boston from tapping the Shawsheen by granting exclusive rights to Cambridge.[51]

Had Boston and Cambridge embraced symbiotic water development, they still might not have gained access to the river. In choosing the Shawsheen, the cities pitted themselves against a powerful foe. The Andover and North Andover industries that were dependent upon the river, like those in the Merrimack River towns of Lawrence and Lowell, wielded significant political power by virtue of their economic importance to their communities and to the state. The extent of industrial development on the lower Shawsheen provided representatives from Andover and North Andover with a convincing counter to the cities' claims regarding the benefits of transferring water to urban areas. Did not the great textile and other mills employ thousands of workers, contribute to state and local taxes and serve the public good? Had not legislation and legal precedent over the last century consistently protected the rights of industries over other Massachusetts water users?[52] Andover responded to Cambridge's and Boston's plans for the Shawsheen with its own petition, asking the General Court to block any outside development of the river.

In 1882, the General Court sought to reconcile the conflicting claims in legislative hearings before the Committee on Public Health. Testimony by representatives from each of the three interested communities revealed the tensions over interlocal relations and water development policy at work during this period. Before the hearings, Cambridge had signed on to Boston's petition for joint development. At the hearings, however, Cambridge withdrew from the agreement, testifying again to the importance of an independent water supply, and insisting that it needed unimpaired access to water revenues in order to offset the costs of other municipal services. Boston alternately cajoled and threatened, arguing that it, too, needed additional water and that the city would oppose any Shawsheen project that did not include Boston.[53]

Andover representatives did not dispute that the metropolitan communities faced a crisis, but instead they focused on their own water needs. They worried that granting water to Cambridge or Boston would decrease flow in the river and thus interfere with industry. They argued that the Shawsheen project represented an unwise change in the use of the river and would open the way for other towns to claim water that Andover needed. Andover challenged the General Court's Committee on Public Health to generate a water distribution policy for Massachusetts. Boston's and Cambridge's proposals for the Shawsheen died in committee, which might seem to be a victory for industrial water users in the Merrimack River Valley and its tributaries.[54] However, the General

Court failed to use this opportunity to articulate a positive water development policy that might have set clearer standards for future water distribution in the region.

A drought in 1883 intensified the crisis. By September 1883, the Boston Water Board estimated that Cochituate held only 45 days' worth of water, and the Mystic only a month's supply. Low water levels concentrated wastes in reservoirs, adding to public agitation. The "steady growth of the industrial interests" in the towns around Cochituate and the new Sudbury reservoirs "so increased the flow of sewage into [Boston's] water sources as to create a very serious and growing evil."[55] The defeat of their Shawsheen proposals, however, had left Boston and Cambridge with no immediate means to remedy these problems.

The Shawsheen episode advanced regionalism by demonstrating yet again the inadequacy of water development plans that depended on interlocal cooperation. Still more important, the defeat turned Boston area communities away from water sources that had substantial existing industrial or urban claimants and impelled them to seek water in rural valleys of less political or economic significance. In California, long experience with just such disputes led to legislative conferences on water policy, explicit laws on the priorities of water distribution, and the creation of state agencies dedicated to resolving water rights disputes. In Massachusetts, by contrast, courts adjudicating water rights and damage claims, created a de facto water policy, which had gradually evolved from one of protecting agricultural rights and commons to one that recognized and assisted industrial development.[56] Thus, cities were bound to come into conflict with industrial interests as they did in the case of the Shawsheen. After the 1882 hearings, Boston and Cambridge did not make that mistake again. In the short run, they concentrated on developing resources to which they had already laid claim. In the long run, they turned to distant, rural water supplies, the development of which required the massing of resources from many communities.

Nashua River

Following the Shawsheen episode, the difficulty of securing new water supplies increased the emphasis that communities placed on water quality and on the control of sewage disposal. Drainage reports issued in 1882, 1884, 1886, and 1888 concluded that the greater Boston area needed a single, coordinated water resources policy. In the 1884 investigation of the Mystic, Charles, and Blackstone Rivers, for example, the General Court explicitly charged consultants to examine both water supplies and sewerage. By the time the General Court approved the Metropolitan Sewerage Commission in 1889, the concept of regional water was also well established. The Shawsheen conflict, taking place relatively early in the transition from municipal to regional sewerage, had thus helped expand regionalism to include waterworks.

Eighteen-ninety-two was the driest year in a decade. The resulting decline in water quantity and quality brought the persistent water problems throughout the metropolitan region to the fore. The Boston Water Board fretted that the city would outgrow the Sudbury watershed in less than ten years. The potential for further water supply expansions seemed grim, as "the needs of the cities and towns neighboring to Boston . . . [would] probably have to be considered" if Boston sought water beyond the Sudbury.[57] Boston petitioned the General Court for relief, requesting specifically that the legislature initiate an investigation of regional water supply. The lawmakers responded, authorizing what was to be the definitive study of metropolitan water resources.[58] The General Court intended that the water question receive "the same general treatment . . . as was adopted by the General Court of 1887 for the creation of a sewerage system . . . ."[59] The State Board of Health embraced the charge fully, determined to quantify existing water problems and future water needs, as well as to assess potential sources for regional development.

The board hired Frederick P. Stearns to carry out the water investigation. Like Chesbrough before him, Stearns had a national reputation as well as a long association with Massachusetts sanitary planning. From 1886 to 1895, he served as chief engineer for the State Board of Health, directing the investigation of the Mystic River and the Charles River sanitation projects. Some years later, Los Angeles hired him as a consultant on the Owens Valley Aqueduct.[60] Also like Chesbrough, Stearns had a clear predilection for large-scale construction. His final report showed little interest in alternatives to regionalism, a bias completely consistent with the State Board of Health's approach to eastern Massachusetts water resources.

Stearns reported his findings in 1894; not surprisingly, they clearly justified a metropolitan water supply. None of the existing waterworks could meet current needs, he contended. Moreover, he predicted that continued urban growth would result in widespread water famine and disease. In fact, typhoid fever was already spreading because so many

Boston area households depended upon heavily polluted supplies. All of the municipal projects constructed in the area, beginning with Boston's first public water "scheme," Stearns reported, had promised abundant water but ultimately failed to meet residents' needs. He acknowledged that these systems fell short because water use "increased beyond all expectation," but still faulted municipalities for these failures. Boston area communities could avoid building a new large supply only by separating drinking water from industrial and wash water. But, because that solution would require laying entirely new pipes and installing new plumbing in all existing structures, it would cost more than simply developing a new supply. On the other hand, a single regional supply, the costs of which would be shared by many towns, would best serve all communities in the greater Boston area.[61]

Although the Nashua River was an obvious favorite from the start, Stearns investigated a number of potential water sources. He rejected the Ipswich, Shawsheen, Charles, and Merrimack Rivers as too small, swampy or polluted for domestic use. He ruled out New Hampshire's Lake Winnipesaukee because of its distance from Boston, the growing development along its shores, and the difficult negotiations required to transport water across state lines.[62]

The Nashua had none of these problems. Stearns found water quality there "entirely satisfactory." Its sparsely settled watershed did not seem likely to attract many new residents or industries. Thus the Nashua supply, Stearns predicted, would be safe from the kind of urban growth that had fouled Lake Cochituate, the Mystic River, and so many other urban reservoirs. Stearns further noted that the Nashua project could easily incorporate Boston's and Cambridge's existing reservoirs, which would significantly reduce the costs associated with building a regional distribution network. Finally, a reservoir on the Nashua would facilitate future expansions into the Ware, Swift, and eventually Deerfield Rivers. Because so many advocates of regional water had criticized Boston area municipal waterworks as shortsighted, the possibility of future growth—realized by the completion of the Quabbin Reservoir in 1939—made the Nashua project particularly appealing.[63]

Following publication of the Stearns report in 1895, the General Court's Committees on Metropolitan Affairs and Water Supply held a series of joint hearings on the proposed Nashua River reservoir and metropolitan water system. In these hearings, representatives from nearly all the towns in the greater Boston area as well as those in the Nashua Valley

The Shawsheen, Nashua, and Swift rivers. Engineer Frederick P. Stearns designed the Metropolitan Water Board's Wachusett Reservoir in central Massachusetts. In the 1920s, his plans guided construction of the Quabbin Reservoir.

voiced their opinions. Stearns' plan received mixed reviews in the Boston area. The suburbs that had no other means to secure adequate water welcomed regionalism without hesitation. Communities that had invested heavily in waterworks reacted with less enthusiasm. Cambridge, despite its earlier pleas for the Shawsheen, insisted that it did not need metropolitan water. Newton joined Cambridge in arguing that they should not have to join the district until their water ran out. Boston, as usual, fell into a category all its own. Initially, the city opposed the district for many of the same reasons that Cambridge and Newton cited. Eventually, however, Boston's leaders came to regard regionalism as a far better solution to persistent water problems than was further independent water development. As was the case with most public works, each town administration balanced expected benefits against political and financial costs.

The legacy of interlocal rivalry and competition surfaced in lengthy debates over the fairness of the metropolitan waterworks. Clearly, Boston area communities saw regional waterworks and sewerage quite differently despite the fact that the two proposals had identical implications for municipal governance. Even the most fiercely independent cities applauded the Metropolitan Sewerage District in 1889, but few cities embraced Stearns' water plans with similar enthusiasm. Some communities that endorsed the regional waterworks feared that another town might reap disproportionate benefits. For those who opposed regionalism, that fear was a nearly insurmountable barrier. Boston, although perceived by its suburbs as predatory, worried that regionalism would force it to pay for suburban water systems. Meanwhile, Cambridge charged that the water proposal was intended to help Boston to "unload" its failing waterworks.[64] In the same spirit, Arlington's spokesman cautioned, "There are dangers lurking in this bill for certain municipalities that don't exist for the city of Boston and it is possible that Boston is getting the long end of this stick." Medford bitterly echoed his accusation.[65] The fact that the Metropolitan Water Board planned to compensate Boston for reservoirs and water lines incorporated into the regional system, but did not intend to reimburse Cambridge, Chelsea, Everett, or Somerville for municipal systems rendered obsolete by the new waterworks, only confirmed the suburbs' sense of injustice.[66]

The distribution of the actual costs of construction proved just as controversial as compensation for the municipal waterworks. Again Boston seemed to reap disproportionate benefits. A regional service district could tax each town according to area, real estate value, or population. Each method would be more onerous for some communities than others. Area-based assessments burdened densely populated Boston and Cambridge less than large, sparsely populated suburban districts. Population- or value-based assessments had the opposite effect, and thus had greater appeal in the suburbs. Because regionalism could not proceed without Boston, the Metropolitan Water Board, like the sewerage agency before it, paid close attention to Boston's preferences in these matters.[67]

In fact, questions of fairness provoked the first crisis in the Nashua project. During the General Court's hearings, several towns refused to join the district citing burdensome taxation and unfair compensation for municipal waterworks and lost water revenues. At various times Arlington, Brookline, Cambridge, Canton, Lexington, Lynn, Medford, Milton, Nahant, Newton, Quincy, Saugus, Stoneham, Swampscott, Waltham, Winchester, and Woburn all objected to the Nashua project for financial reasons.

According to the State Board of Health's own estimates, only Brookline, Cambridge, Waltham, and Winchester had no immediate need for new water supplies. Their opposition to metropolitan water, therefore, surprised no one. For these towns, regionalism represented an unnecessary expense and a threat to municipal independence. A number of the towns that opposed regionalism, however, had no reliable water supplies. Lexington, Lynn, Nahant, Newton, Saugus, and Woburn had outgrown their waterworks by the 1890s. Upstream growth contaminated Stoneham's supplies.[68] Nevertheless, in these communities, a desire to maintain political independence inspired an overly optimistic assessment of their water systems.

The question of fairness was raised most explicitly by cities like Arlington and Medford, which admitted that they needed regional water. Before the General Court, Medford's representative argued that each city should fend for itself. "If a city gets in a hard pinch, it is for itself to get out of it if it can," he said.[69] In this case, Medford's objections seemed to be a veiled bid for taxation and compensation arrangements to benefit Medford residents.

Unhappy about the financial details or unconvinced that they needed regionalism, many communities sought to exclude themselves from the district. So many towns petitioned to withdraw that the regional water plan seemed in danger of collapse. Everett, Somerville, and several other towns feared that granting so many exemptions would raise the cost of regionalism beyond their means. They urged the water board to make district membership compulsory. These communities proved less concerned about the fairness of the system than about keeping the waterworks affordable for all towns. To assuage these fears about the project's price tag, the final legislation exempted Arlington, Brookline, Cambridge, Quincy, Milton, and Stoneham from the district, but required them to pay their portion of construction costs when they did join.[70]

Boston had its own suspicions that Stearns' plan would burden it unfairly. Although Stearns' proposal to connect the Nashua aqueduct to the Sudbury-Cochituate distribution pipes made the best possible use of existing infrastructure, it was to cost Boston two million dollars in water revenues annually. Boston city officials resented this proposed appropriation of their water system. Alderman Martin Lomasney accused the Metropolitan Water Board of taking Boston's water and taxes to supply outlying towns.[71] City councilors reacted by instructing Mayor Edwin Curtis to "take all steps necessary" to assess the constitutionality of the metropolitan water act.[72] The Water Board sought to mollify Boston with nearly

Wachusett Reservoir, completed in 1906. Courtesy of the Massachusetts Archives, Boston, Massachusetts.

Wachusett Dam under construction in 1902. Courtesy of the Metropolitan District Commission Archives, Boston, Massachusetts.

Cutting stone for Wachusett Dam with a pneumatic drill, 1902. Courtesy of the Metropolitan District Commission Archives, Boston, Massachusetts.

Because water filtration systems were neither perfected nor completely trusted, the Metropolitan Water Board ordered all topsoil stripped from the Wachusett Reservoir site. Courtesy of the Metropolitan District Commission Archives, Boston, ssachusetts.

Metropolitan Water Board engineers, 1897. Courtesy of the Metropolitan District Commission Archives, Boston, Massachusetts.

Immigrants hired by labor contractors lived in hovels near the Wachusett Dam site. Although their living conditions did inspire some sympathy, more frequently the fact that immigrant laborers lived in such squalor added to local townspeople's mistrust of their new neighbors. Courtesy of the Metropolitan District Commission Archives, Boston, Massachusetts.

fourteen million dollars in damage payments which, of course, aroused further jealousy in other communities.[73] Ultimately, however, the need for improved water supplies and the impossibility of proceeding alone overcame most of the political and financial objections to metropolitan water.

Opposition in the Nashua Valley

If compromise and compensation finally persuaded most of the communities in the Boston metropolitan area to back regional water, they did little for the people living near the proposed reservoir site. The Nashua River reservoir, eventually named after nearby Mount Wachusett, was to inundate twelve square miles of land in three Nashua Valley towns: Boylston, Clinton, and West Boylston. Clinton would lose nearly a third of its developed land and water power for several of its local mills. In West Boylston, only the farms lying outside of the village center, which represented almost forty percent of the town's taxable lands, would remain above the flood. Boylston, meanwhile, would lose all but its town center, and more than forty percent of Boylston residents would lose their homes.[74] Cut off from their markets and deprived of farming and manufacturing incomes, these communities faced an uncertain future.

Metropolitan Boston's water claims struck many central Massachusetts observers as illegitimate. That water disputes were, in the words of the *Clinton Daily Item* editor, "as a general thing . . . decided in favor of the city, on the specious grounds that a water supply for cities is a necessity while the country towns can get along without it" merely heightened the small towns' suspicion of cities and of the Massachusetts legislature. Of course, Boston was not the only community taking rural water. In the 1890s, Springfield, Holyoke, Northampton, and Chicopee all "seize[d] the property of their rural neighbors."[75] But even in Worcester, a sizable city with water problems of its own, newspapers criticized Boston's apparent tendency to view ". . . all the natural water courses and ponds in Massachusetts as intended by the Creator of the universe for its own special and absolute use whenever its sweet will and pleasure so select."[76] These observations point to two major problems created by late-nineteenth- and early-twentieth-century resource policy. First, the transfer of water from rural to urban uses sacrificed rural prosperity and independence for urban, industrial growth. This reinforced antipathy towards urban areas, as illustrated by the Nashua Valley's opposition to the Metropolitan Water District. The second problem was that the movement of water across tremendous distances greatly extended the dimensions of resource competition, and brought cities, and eventually states, into competition over water. The battles arising from interbasin transfer are usually associated with twentieth-century development in the American west, but such transfers were in fact pioneered in Massachusetts and New York.

When faced with Stearns' proposal to export their water to the Boston area, the Nashua Valley towns did not put up a very effective struggle. They did not seize upon the objections of larger communities like Worcester or Springfield or directly attack Massachusetts' pro-urban, pro-industrial water policy. Instead, they fretted over the direct damage that they expected from the Wachusett plan. Representatives from Boylston, Clinton, and West Boylston did seek alliances with a few potential opponents to the project, including Boston and Nashua, New Hampshire. Boston, however, merely proposed a delay, and then decided to support the plan. Because the Metropolitan Water Board planned to divert only floodwater into its reservoir, the project never posed a serious threat to mill owners in Nashua. They, therefore, raised no serious objections to the plan either.

Protections for the Nashua mills were quite deliberate. Similar provisions also ensured that the manufacturing center of the Nashua Valley, Clinton, would not suffer unduly. In Clinton, the Metropolitan Water Board agreed to construct a new steam power plant for the Lancaster Mills to replace lost water power.[77] A few much smaller mills were to be destroyed by the reservoir, but compensation promised to the owners and employees overcame their objections. Stearns had clearly learned the lesson of the Shawsheen River. Not only did he target relatively weak communities that existed on the margins of the Massachusetts economy and political system, but he ensured that industrial interests would not raise significant objections. By protecting the Metropolitan Water Board from attack by industrial interests, Stearns provided the General Court with a simple economic calculus. Sacrificing farms and houses but not significant industrial development seemed a reasonable price to pay for improved public health and continued industrial growth in the Boston area.

The protections offered to Wachusett's potential industrial opponents derailed the source towns' fitful efforts to block the reservoir. After they failed to find an ally in Boston or Nashua, the towns readily admitted defeat. Boylston, Clinton, and West Boylston representatives urged the state to finalize regional plans quickly so as to allow local residents to get on with their lives.[78] In their despairing, ethnically charged request that "the robed priests . . . dispatch us with as much humanity as they possibly can" can be heard the rural communities' sense of the city as threatening, foreign, and omnipotent.[79]

If Boylston, Clinton, and West Boylston conceded defeat early in the metropolitan water battle, they had good reason. By 1895, none of the three towns could absorb any more economic dislocation. The 1893 depression had devastated Clinton's major employer, the Lancaster Mills. At the time, the company employed a fifth of Clinton's population. When the mill put all 2300 employees on short weeks, the impact rippled through the entire community. A year later, a canvass of West Boylston residents revealed that 2189 of the towns 3019 inhabitants expected to leave their community "on account of loss of employment."[80] Dispirited and burdened with high unemployment, these communities had few resources to mount an effective campaign to stop the proposed reservoir.

Of course, the reservoir compounded economic problems in the Nashua Valley. When the Wachusett Reservoir claimed establishments in the Boylston village of Sawyers Mills and the F. H. Rice cotton mill in West Boylston, unemployment jumped in all three towns.[81] Rising waters forced the Central Massachusetts Railroad to move its tracks from Boylston to Clinton. The *Clinton Daily Item* described the expected fifty-two trains passing through Clinton as an "annoyance to business interests" and a hazard likely to interfere with local street transportation.[82] Meanwhile, the newspapers were filled with lists of people who had sold off their lands to the water board as well as with tales of the sheriff's efforts to evict families who refused to accept compensation. One such report described four old women who had to be carried to their new residence.[83] While waiting for the water board to take their homes, Nashua Valley residents let their fields lie fallow and ignored their businesses. Many reported making no provisions for the future, for fear that they would waste their effort.[84]

Once the reservoir construction began, the project attracted thousands of newcomers to the Nashua River Valley. Newspapers told of women, wooed, wed, and bilked of their savings by "clever crooks" who gained entry into local society by first finding employment with the district.[85] Land speculators also arrived, buying up mill privileges—rights to specific amounts of river water to drive industrial water wheels—and acreage in the proposed reservoir basin. They were able to sell their newly acquired property at inflated prices because the Metropolitan Water Board was desperate to secure title to the construction zone.[86] Much as valley residents resented the speculators and other impostors, it was the presence of immigrant laborers and the contractors who hired them that inspired the greatest antipathy.

The Metropolitan Water Board did not hire Wachusett Reservoir workers directly but relied on contractors to supply workers. Instead of hiring locals, these contractors, called *padrones,* brought immigrants, especially from Italy, into the small Nashua Valley towns. *Padrones* not only controlled District hiring but also required employees to purchase supplies from the *padrones'* commissaries and rent the *padrones'* squalid shacks.[87] Thus, they shut out Clinton, Boylston, and West Boylston merchants from the benefits of the construction project and made reservoir employment less attractive to local residents. Although the metropolitan water act specified that "preference in employment shall be given to citizens of . . . [the] Commonwealth," only five percent of the laborers on the Wachusett Project were listed as Massachusetts citizens.[88] In the context of the Nashua Valley's economic depression, the exclusion of local laborers was devastating.

Clinton's largest labor union objected to the contract labor system on the grounds that *padrones* brought "a very undesirable class of help into the town" through their "influence upon the ignorant Italians and Hungarians" hired for "little or nothing . . . ."[89] Toward the end of construction, the reservoir project was delayed by several strikes over wages and commissary purchases, which, instead of cheering the union, merely made the laborers seem unruly.[90] Here, the cities were held responsible not only for the loss of jobs and land, but also for the importation of the immigrant poor, already widely seen as a dangerous, corrupt, unhealthy, and immoral influence on politics and society.

Lurid descriptions of knife fights, murder, and drunkenness, such as the stabbing of Lorato Demarzio in a fight over a round of drinks, reinforced the negative image of the immigrants.[91] Newspapers trumpeted stories of disagreements over shanty businesses that escalated into charges of theft, and of numerous violations of local liquor prohibitions.[92] Alcohol consumption was one of the factors that most discredited immigrant workers. The *Clinton Daily Item* was particularly critical of the utter disregard for local liquor laws by laborers and their *padrones*. Because the Water District contracted with the *padrones* in the first place, it was perceived as condoning the illegal sale of alcohol.[93] According to Clinton's largest union, labor contractors were nothing more than big city liquor dealers, who supplied their employees with beer and tobacco instead of paying them sufficient wages. [94] This allegation, while emphasizing morality issues, also reflects the Nashua Valley's continued anger at Boston.

In January 1900, responding to his furious constituents, Clinton's legislative representative, D. I. Walsh, requested an investigation of the Metropolitan Water Board's activities in central Massachusetts. He accused the board of wasting public money and violating state labor and liquor laws. He was particularly distressed that *padrones* sold alcohol without liquor licenses and refused to hire Massachusetts citizens or pay their laborers weekly. In the General Court, representatives' reactions to the proposed investigation corresponded to their proximity to the reservoir and their relations with Boston. For example, Worcester's representatives sympathized with Clinton, "a community in whose domain a horde of ignorant foreigners had squatted." They accused state agencies of "sucking the life blood from the very 'heart' of the commonwealth."[95] In contrast, legislators from towns closer to Boston dismissed the proposal as a transparent effort to force the District to compensate Clinton more generously.[96] Debate over the investigation order revealed the factions that had by this time divided the state. On the one side were those in favor of local resource control; on the other, those who sought centralization and, with it, large-scale public works construction and greater professionalism.

Resentment persisted for years in the Nashua Valley. After the Metropolitan Water Board completed Wachusett Reservoir, a representative of Clinton's business community, Charles Choate, summed up the community's objections. The regional water system, he lamented, had severely undercut the political independence of his town and nearby communities. In constructing the reservoir, the water agency had forcibly taken water and land for the sole use of a distant city. The water commissioners, Choate concluded, had far too much unrestricted power:

> No other public servants in this Commonwealth are given so large a scope of power, or so enormous a sum of money to expend. No other public work so directly and inevitably involves injuring one portion of the State for the sole and exclusive benefit of another. No other public board . . . has ever existed . . . in whose personality the interests and influences of one portion of the community, and that the one to be benefited, have so exclusively prevailed over the interests and influences of another portion of the community, and that the portion to be injured.[97]

In this statement, Choate encapsulated the transformations wrought by regional public works. Regional agencies wielded expansive new powers over natural resources in both source and consumer communities with very little input from either. Regionalism, moreover, altered the relationship between source regions and urban centers. Special districts subjected the needs of sparsely populated areas to those of the urban metropolis. In essence, metropolitan services ended the conflicts over resources and power that had fragmented the metropolitan area. But in the process, they divided urban and rural interests and clearly sacrificed small towns and agriculture for large cities with their industry, immigrant populations, and powerful institutions.

Implications of Boston Regionalism

By 1895, the transition to regionalism in metropolitan Boston was complete. Metropolitan boards oversaw both sewerage and water supply.

Applied to an ever wider array of services, special districts offered an ideal balance of service, centralized administration, financial flexibility, and municipal independence. Regional agencies were themselves at once familiar and innovative. They closely resembled the independent municipal water boards that had overseen Boston's waterworks for half a century. For decades the engineers who designed regional systems had been honing their skills on municipal systems for Boston, Cambridge, and other communities. In the case of sewerage, in particular, the regional plan replicated designs drawn up as part of municipal sanitation studies. Nevertheless, the Metropolitan Water Board and Metropolitan Sewerage District significantly redistributed political power in the Boston area. Cities remained at least nominally in control of their public works; they retained their water and sewer main and branch lines. However, they lost authority over new construction, major distribution networks, resource development policy and—all important to local politics—hiring on public works projects. The new agencies institutionalized the growing power of professionals and experts. Furthermore, they established new enclaves of power for the Yankee elite pushed from local government by Boston's growing immigrant minority. Significantly, this transition took place, by and large, with the acquiescence and even enthusiasm of the very municipal officials who stood to lose the most with the transfer of public works out of local hands.

Municipal departments, headed by elected or appointed officials, had been highly responsive to service requests from individuals. Regional works, in contrast, operated independently of state or local oversight, and therefore at some remove from voters. Throughout the nineteenth century, this distance was justified as the means to ensure scientific and apolitical public works development. The new institutions, controlled by engineers and bureaucrats, developed their own priorities and alliances that might or might not have best served the mass of residents.[98] People supported this reorganization because increasingly complex technology seemed to require specialized knowledge, and because they faced pollution and water supply problems that their municipalities had repeatedly failed to solve.

The popularity of improved water supply and sewerage overcame the political infighting and fragmentation that usually defeated political reform efforts. Widespread support for improved services did not, of course, entirely erase deep-seated divisions in Boston and Massachusetts. Throughout the 1870s and 1880s, Republican leaders had advocated annexation and administrative reorganization to contain Democratic—and by extension immigrant—power. Some of the city council's own reports explicitly promoted annexation as a means to halt corruption by diluting working-class and immigrant votes. In the same vein, Brookline and Chelsea rejected annexation for fear that the influence of Boston's poor wards would infect their politics.[99] Within the State House, Republicans dominated both the Water Supply Committee and the Metropolitan Affairs Committee when regionalism came before the General Court.[100] But the changes wrought by regionalism did not affect only Democratic-controlled communities. Most of the towns included in the regional district were dominated by Republicans. Moreover, the three towns that lost the most from regional water, Boylston, Clinton, and West Boylston, were also solidly Republican communities. While perhaps intended to contain the power of Boston Democrats, regionalism sacrificed the political and economic independence of many Republican towns to the needs of urban Democrats.

True, the regional water and sewer systems allowed the Metropolitan Water Board, identified with Republican reformers, to appropriate Boston's water system at a time of increasing partisan competition in municipal politics. But both the new water and sewerage commissions also eliminated the most important tasks of municipal board members who were already sympathetic with regionalism and gave city officials the boon of being able to improve public services without raising taxes. Ironically, the new agencies, and the Water Board in particular, interfered more with the political power of potential Republican allies in central and western Massachusetts towns and in Boston's suburbs than they did in Boston itself.

Although the coalition behind regionalism blurred partisanship, negative images of urban squalor in general—and in Boston's immigrant neighborhoods specifically—reinforced other political divisions. Specifically, moral-environmental theory deeply colored suburban and rural aversion to Boston's looming presence. In late-nineteenth-century cities, reformers drew moral conclusions from the squalid conditions of working-class neighborhoods that extended to the political morality of these neighborhoods' municipal representatives. As negative reactions to Boston's public works development in the Mystic and Nashua Valleys so clearly demonstrated, the image of morally, physically, and politically corrupt urban neighborhoods, together with a good measure of late-nineteenth-century nativism, cast a shadow across the whole city. Antiurban sentiments found even more fertile ground when these negative images were reinforced by the city's threats to local autonomy in suburban and rural communities and by appropriation of their natural resources.

Urban political reformers took full advantage of moral-environmental rhetoric. They used Boston's apparent inability to eliminate sewage from the Mystic and Charles Rivers or to keep pace with water demands to stir resentment of municipal corruption and impotence. As further evidence of corruption, reformers cited ward-based political organizations that, they claimed, thrived on the spoils system. Whether in the State Board of Health, in the General Court or in Boston's city council, leaders who sought to centralize and streamline public administration readily adopted disease metaphors to further their cause. Anticorruption campaigns took place at a time when Boston's immigrant communities had begun to exercise significant influence in municipal politics. At the time, many of Boston's city leaders depended upon the support of immigrant voters who lived in unsanitary districts and had, in the eyes of the elite, questionable morals. This association between municipal politics and the least "proper" of Boston's voters lent credibility to reform rhetoric linking urban physical conditions and political corruption. The application of moral-environmental theory to urban politics reinforced antiurban sentiment both in Boston and beyond its borders. Perceptions of cities as corrupt and foreign would persist into the twentieth century and would fuel similar opposition to similar metropolitan projects in California.

Proposals to centralize public works encountered the greatest opposition when a new system seemed to sacrifice the welfare of small towns for the benefit of a larger city. The more distant the city, and the more important the resources transferred to it, the more bitter the opposition in the smaller community. Regional public works expanded government authority and monopolized water resources for industrial and urban use, a situation that would prove especially important to water-hungry California cities such as Oakland. But this transfer of resources, when implemented through regionalism, revealed an inherent irony of urban reform. It is true that the new agencies balanced home rule and central administration and reduced the access of municipal bosses to public employment, thus reducing the spoils available for distribution to their constituents. At the same time, however, the new institutions sacrificed the economic viability and independence of rural towns for the benefit of the metropolis.

By designing sanitation and water supply for whole watersheds, the metropolitan districts diverted wastes away from population centers and protected drinking water. The improvements were particularly noticeable in suburban towns that could neither compete with Boston for water resources nor afford to develop independent waterworks. Intercepting sewers finally diverted wastes away from the congested urban waterfront. But, unfortunately, these systems would eventually pose serious threats to Massachusetts rivers. As metropolitan communities replaced local with distant water supplies, they not only reduced rural communities' autonomy, but also freed themselves from having to clean urban streams. In the worst cases, industrial rivers such as the Merrimack were exempted from most water quality regulations and thus remained hopelessly contaminated.[101] Moreover, efficient though they were, regional sewers did not prevent untreated wastes from overflowing into rivers and streams whenever rain overfilled intercepting sewers. Some of these problems would not be identified for many decades. In the meantime, regional sewerage and water both improved urban sanitary conditions and eliminated the water shortages that threatened growth and public health throughout the Boston area.

4 | The East Bay: Regionalism in the Progressive Era

At the end of the nineteenth century, while Boston celebrated a successful transition from local to regional water and sewerage administration, private companies continued to dominate public services in California. In the East Bay, water companies maintained their autonomy despite municipal efforts to assert greater government control over pricing and service. Meanwhile a series of mergers, which concentrated East Bay water resources in fewer hands, solidified the private companies' monopoly and increased their ability to resist regulatory initiatives.[1] Corporate consolidation did not improve water service, however. Customers continued to complain about shortages, high water bills, and inadequate fire protection. By 1919, public dissatisfaction with the service offered by private companies had come to resemble the dissatisfaction Boston area communities had experienced with municipal water and sewerage networks several decades earlier. But, because California communities lacked Massachusetts' tradition of public resource ownership and development, regionalism in the East Bay communities of Berkeley, Oakland, and Richmond depended first and foremost on curbing corporate power.

Progressives led the campaigns for regionalism in California. They had had their start as urban reformers intent on weeding out machine politics. The term "machine politics" in California, however, referred as much to railroad and utility influence, as to the working-class, ward-based power networks that urban reformers battled back East. In fact, corporations had been a favorite political scapegoat since the 1870s; Californians commonly blamed the railroad "octopus" for economic as well as political frustrations.[2] The Progressives rode this familiar rhetoric to electoral and political success beginning with Hiram Johnson's 1911 gubernatorial election. By 1915, what had begun as a fairly narrow attack on machine politics, had emerged as a comprehensive effort to exert public authority over a variety of monopolies, including water companies, that were charged with corrupting local and state politics and impeding economic development.

From 1914, the Bay Area was a stronghold of the Progressive Party.[3] Familiar antirailroad rhetoric was applied to the water companies, and agitation for regulation and public ownership soon followed. Although no direct ties existed between the East Bay's water and railroad companies, advocates of public ownership consistently linked the two utilities. The *Oakland Press* gave voice to common suspicions when it described a water company merger as taking place "under the parental wing and guiding influence of the Southern Pacific."[4] Similarly, the Progressive weekly *Pacific Municipalities* associated all private utilities with tyranny, unrestrained corruption, and the demise of municipal democracy.[5] Where Eastern reformers blamed municipal corruption on immigrant-supported machine politics, California Progressives targeted the railroad and private utilities as the source of all political evil. Ironically, their partisan opponents in California's Democratic Party had attacked the railroad monopolies for decades before the Progressives emerged as a significant political force.[6] Anti-utility sentiment in the East Bay played the role that antiurban feelings had played in galvanizing public support for reform programs in Boston.

In the East Bay, regionalism had its roots not only in long-standing suspicion of utilities and Progressive efforts to curb water company power, but also in dissatisfaction with local water supplies. By the 1910s, East Bay residents commonly complained about badly designed distribution networks, poor service, low quality, high rates, and above all, impending water shortages. The shortcomings of the water network and the ever-present threat of drought created a series of real and perceived water crises that mobilized local leaders and voters to demand changes in the East Bay's water system. Between 1880 and 1915, these demands precipitated so many public water proposals that the subject became a permanent feature of East Bay political discourse. In fact, improved water supply came to be promoted as a panacea for the East Bay's economic and political woes.

Early Regional Initiatives

By the early 1900s, the East Bay cities had outgrown the creeks and the wells that fed their water systems.[7] After 1910, many observers were convinced that narrow water mains and reservoirs on the Temescal, San Pablo, San Leandro, and Pinole Creeks, even supplemented with wells in Alavarado, failed to provide enough water to permit economic development. When the water companies raised their prices despite consistently substandard service, agitation for public waterworks gathered momentum. Facing similar conditions, San Francisco and Los Angeles built municipal networks which have successfully served ever-growing populations.[8] In the East Bay, however, there was no clear consensus on how to proceed. Some advocated municipal development and city-county consolidation, and others symbiotic construction. Still others spoke out for a regional approach. Although past experiences with public water initiatives had demonstrated the pitfalls of single-city projects, regionalism still appeared too threatening to gain widespread public support.

Interlocal ventures were no more popular than regionalism. Between Oakland's rejection of the Bay Cities Water Company proposal in 1905 and the final approval of the regional East Bay Municipal Utility District (EBMUD) in 1923, three major multiple-city water schemes went down to defeat. As in Boston, East Bay communities jealously protected their autonomy from each other and from possible consolidation with San Francisco.

Two other major obstacles stood in the way of public water development. First, none of the East Bay public water plans proposed during this period identified a viable source of water, yet without water no public development could proceed. Second, opposition to taxes and government enterprise had deep roots in East Bay politics and served to reinforce the private water companies' tenacious opposition to public development. The absence of a clear mandate for public, as opposed to private, services delayed regionalism in the East Bay despite the persistent demand for improved water supply.

Municipal Water District Act of 1911

The story of the East Bay's first regional water proposal reflects Californians' ambivalence towards public enterprise. From 1909 to 1914, Oakland's Progressive mayor, Frank K. Mott, and Berkeley's Socialist mayor, J. Stitt Wilson, championed a metropolitan water initiative that would have included much of Alameda County. In taking this approach, Mott acknowledged not only perennial East Bay water problems but also the financial and political barriers to municipal development that had appeared in earlier public water campaigns. Unfortunately, Mott took on urban water supply at the same time that Progressives in Sacramento were endeavoring to centralize utility regulations and water resource policy. Their efforts threatened municipal authority, and so complicated the question of public ownership that support for public waterworks flagged.

If Mott and Wilson had needed only a water emergency to convince voters to approve their plan, they could hardly have chosen a more promising decade. Thousands of San Franciscans fled the devastation of the 1906 earthquake and fire to settle in the East Bay. As in Boston, population growth and increased water consumption forced water companies to expand their use of even contaminated water supplies. By 1907, the Oakland Board of Health judged the Contra Costa Water Company's supplies too polluted for use in public schools.[9] In 1909, polluted surface runoff contaminated wells owned by People's Water Company. More than a third of Alamedans suffered gastroenteritis as a result.[10] The epidemic cast a pall over People's Water Company plans to build three new reservoirs close to other sources of urban pollution. Such incidents confirmed the unreliability of both private water development and local water supplies.

The district that Mott and Wilson proposed was to include only the Alameda County communities of Oakland, Berkeley, Alameda, Albany, Emeryville, Piedmont, and San Leandro. The mayors clearly understood that projects of the scope built by Los Angeles and San Francisco were beyond the reach of their smaller communities. The boundaries of the district also reflected the results of water company mergers that had already created a privately owned regional system in the East Bay.

As in Boston, East Bay regionalism had obvious advantages over municipal construction. A regional water district increased the financial resources available for water development and skirted state-imposed spending caps. According to backers, "The district plan enable[d] the carrying out of a water supply project which [was] financially impossible for any one city within the proposed district" and did "not impair the bonding capacity of any city."[11] Mott and Wilson remembered that competing demands for public funding had helped defeat Oakland's 1905 purchase of Alameda Creek water rights from the Bay Cities Water Company. By removing water bonds from the general city budget, a special district freed city leaders from having to choose among popular public projects. Mott and Wilson anticipated other benefits from regional administration as well. Cities could only issue bonds secured by the value of local property, but public works districts could issue revenue bonds, borrowing against future earnings and thus freeing enormous sums of money for public works construction.[12]

After agreeing upon a general approach to regional water, Mott, Wilson, and the Alameda County mayors who supported their efforts sent a proposed bill to the state legislature. The bill passed easily. Signed by Governor Hiram Johnson, the Metropolitan Water District Act of 1911 authorized communities around Oakland to join in a Metropolitan Municipal Water District, pending voter approval.[13] Despite the growing problem of water contamination, the bill received mixed reviews in the East Bay. Of course, People's Water opposed the measure. But prominent Progressives also expressed serious reservations. As a result, efforts to implement the 1911 legislation became tangled not only in the difficult question of public versus private ownership, but also in the issues of home rule and state control of natural resources.

In fact, the East Bay's first regional water proposal coincided with a concerted effort by Progressives to increase state supervision of both private utilities and natural resources. Shortly before Mott and Wilson sent the metropolitan water bill to the legislature, John M. Eshleman and other Progressives in Sacramento sponsored the Public Utilities Act of 1911. Eshleman's measure extended the regulatory authority of the California Railroad Commission to all private utilities in the state. Incidentally, this measure created concern that the Railroad Commission would interfere with local autonomy by attempting to regulate publicly owned municipal and metropolitan projects as well as urban private utilities.[14] That same year, water reform legislation created a commission to investigate water rights, declared all water to be public property, and limited hydroelectric power permits to twenty-five-year terms. This measure, like the Public Utilities Act, reflected growing fear of utility monopolies and frustration with speculative water claims. However, instead of freeing water for productive use, the new statute created a rush of water claims, threatening to severely limit opportunities for municipalities to build waterworks, regardless of how hard-pressed for clean water their residents were.[15]

Progressives throughout California celebrated the Railroad Commission's new powers, and not just because they promised to curb corporate power. Eshleman himself promoted state regulation as the answer to the graft that accompanied cities' attempts to regulate powerful utilities.[16] Because they preferred state to local administration, the utility companies backed Eshleman's reforms. They saw state supervision as a means to eliminate damaging competition and problems that arose when utility networks crossed city lines. They argued that municipal politics interfered with scientific management of utilities, that city workers lacked proper incentives to work hard, and that city officials did not know enough about technical matters to regulate private utilities effectively.[17] Furthermore, they hoped that regulation would silence demands for municipal ownership of water, power, and transportation networks.[18]

The fact that Pacific Gas and Electric and other major utilities welcomed Eshleman's reforms made East Bay leaders suspicious. Even though the final draft of the bill limited state oversight of municipal utilities, defenders of municipal governance argued that regulation was a critical municipal right and a guarantor of local autonomy, particularly where city-owned utilities were concerned.[19]

By 1911, regionalism itself was not particularly controversial. Semiautonomous districts were a well-established feature of California governance and, in the words of one engineer, "no untried experiment." In support of the regional water proposal, proponents also could point to the success of Boston's Metropolitan Water District, as well as the problems with their own waterworks.[20] Nevertheless, in fiercely independent Oakland, the financial and service benefits of the Mott-Wilson Plan could not overcome public resistance to either creating a new governmental authority or opening municipal affairs to state government meddling. When the district came before voters in 1914, Berkeley approved the proposal, but Oakland defeated it, voting two to one against the measure.[21]

The water company had lobbied hard against the Mott-Wilson Plan. William Dingee, former president of the Contra Costa Water Company assured Frank Havens of People's Water Company that the utility had "thoroughly satisfied every intelligent man, woman and child in the vicinity of Oakland of the utter futility of the Municipality running a water plant in competition with an existing privately owned plant." Furthermore, because Oakland had no ready water supply, the threat of competition was an empty one.[22] But the dearth of water was not the only problem for the Mott-Wilson plan. Ironically, Progressives themselves indirectly helped defeat the East Bay proposal. By using the Public Utilities Act of 1911 to attack corruption in California cities, Progressives in Sacramento had blurred established notions of the threats to municipal democracy. Before the state stepped in, both Democrats and Republicans had commonly portrayed the corrupting influence of utilities as the greatest threat to elected government. But, as reformers achieved increased state supervision of municipal administration, defenders of city governance felt the ground shift under their feet. They still saw utilities as dangerous, but now state regulations also threatened municipal authority. They would not endorse limits on home rule even in the name of increased public control over resources and monopolies. Their only alternative was to support municipal regulation of existing utilities, and implicitly, the continued private ownership of East Bay water. In this case, issues of local autonomy and public authority over utilities and services became tangled. Although the public still clearly wanted some kind of government authority regulating utilities and resources, there was a striking lack of consensus over what shape that authority should take.

Richmond Water Commission

In 1916, People's Water Company proposed a new dam on San Pablo Creek. Located close to urban development, and therefore likely to become contaminated, the San Pablo project met with considerable opposition, even though the East Bay clearly needed new water supplies. North of Berkeley, Richmond and several other Contra Costa County communities reacted to the proposed dam with public water development plans of their own.[23] Public ownership advocates in Richmond used a slightly different approach than had Mott and Wilson in Alameda County. In 1916 Richmond voters approved a referendum that established a water commission and envisioned a regional network for Richmond, El Cerrito, Giant, Stege, and San Pablo. However, this referendum did not spell out a specific plan for water development. Nevertheless the Richmond Water Commission, although intended as a governmental institution to champion public water, was to prove just as vulnerable to utility lobbying and public ambivalence as had the proposed water district in Oakland and Berkeley.

The Richmond Water Commission set to work immediately upon its formation. After thorough study, the commission rejected East Bay water supplies as inadequate and instead proposed that the Contra Costa communities purchase ninety million gallons of water a year from the Marin Municipal Water District. Even at twenty thousand dollars a year plus transportation fees, the Commission estimated that Marin water would cost consumers fifteen cents per unit less than what People's Water Company charged. The promise of lower water rates and public ownership earned the Richmond Water Commission much acclaim.[24] But, as Oakland and Berkeley public water advocates had discovered in 1914, public dissatisfaction with private utilities could not yet overcome entrenched resistance to government growth.

Almost immediately upon publication of the Marin plan, opponents sought a city referendum to dissolve the Commission. The Richmond Taxpayers Association led the movement, arguing that the Richmond Water Commission held dangerous powers, including an "unlimited power to levy taxes" without voter approval.[25] Richmond's Central Labor Council, despite its declared support for public ownership of utilities, sided with the Taxpayers Association on this issue. Richmond's city charter, unlike Oakland's, permitted municipal water development. The Labor Council argued that exercising this authority would permit public ownership without raising taxes as much as supporting the Richmond Water Commission would.[26]

Proponents of public water accused People's Water Company of setting up the Richmond Taxpayers Association as a front to oppose public ownership and maintain its monopoly. Prominent civic organizations, including the Richmond Club and the women's clubs that had long championed urban improvements, saw the Taxpayers Association's actions as evidence of corporate interference in municipal governance.[27] The fact that the Taxpayers Association kept its membership secret did little to assuage these suspicions. Eventually, the *Richmond News* revealed that the president of the Belt Line Railroad also led the Taxpayers Association. While not direct evidence of water company meddling, this information indicated that local utilities were covertly lobbying against public ownership.[28]

Revelations about the links between the Taxpayers Association, public water opponents, and utility companies were not enough to protect public water, however. In 1917 Richmond voters abolished the year-old Water Commission. The *Richmond News* greeted this event with a bitter editorial that emphasized the voters' gullibility and the power of the water company:

> The sovereign people have said by their votes that the water district shall be disincorporated. The Water Trust wins. Long live the King! . . .
>
> We believe the people have made a serious mistake in their choice, and will soon come to know and realize it. They have taken a step backward which it will take years to rectify. All right. The people have spoken. Long live the people! . . .
>
> We still maintain that municipal ownership is coming so fast, all over the United States, that nothing can head it off. Richmond in time will come to

> it and adopt it, as all other progressive cities are, but when that will be, or how, Lord knows now—we don't.
>
> And as we said at the start of the campaign, we intend to put in our own well and waterworks next spring and bid the water corporation personal defiance. And everybody else has that same privilege if they wish to exercise it.
>
> And now, having had nothing but water for some time past, we believe we will try a cocktail. Dry Manhattan, please—no cherry![29]

This editorial reflects continued frustration both at the unwillingness of voters to hold firm in their support of public ownership and in the power of the water companies to manipulate local politics. Ultimately, advocates of regionalism believed, the East Bay could eradicate the water companies' corrupting influence only by eliminating private water monopoly. This position was, of course, analogous to the one held by railroad reformers in Sacramento, and to the strategy that proponents of regionalism in Boston used to tackle corruption in their city by removing public works from municipal control.

Public Utilities District Act of 1915

While Richmond was creating and dissolving its Water Commission, Alameda County cities continued their own struggle to improve local water service. At the same time, problems with water supply persisted throughout the East Bay. In 1914, People's Water Company defaulted on its debt. New reservoir construction stalled. By 1915, public water advocates had introduced a new bill, which was approved as the 1915 Public Utilities District Act. Boosters, including sixty representatives of "civic and improvement associations" not only contributed to the legislative success of this proposal, but kept the issue of regional water development on the public agenda.[30] Even so, they could not forge a consensus on how to implement public water development or where to get the water that the East Bay needed. The water debates that took place between 1915 and 1918 demonstrated more clearly than ever the political conflicts that delayed public ownership in the East Bay.

The 1915 Public Utilities District Act did not immediately yield a concrete water plan. As in 1911, the proponents of public water seemed to exhaust their energy and political capital just getting legislation passed. Finally, however, in 1918 mayors and representatives from Oakland, Berkeley, Alameda, and Richmond convened to discuss water. The group initially focused on Hetch Hetchy. San Francisco's project was tantalizing. The East Bay could not identify a comparable water source and, moreover, cooperating with San Francisco seemed much more efficient than building a separate, parallel network. Cooperation with San Francisco was extremely controversial because it seemed to threaten East Bay autonomy. So, the meeting quickly turned to discussions of other ways to implement the 1915 Public Utilities District Act.[31] As a result of these meetings, representatives of the East Bay cities concluded that their communities should form a district to furnish water and other unspecified services throughout Contra Costa and Alameda counties. This plan was far more comprehensive than the metropolitan water district proposal defeated in 1914. Not only would the new utility district cross county lines, but it would also pave the way for public ownership of a wide range of services.

The utility district idea received considerable support from the East Bay Public Ownership League. The league was affiliated with a national organization dedicated to "protecting and promoting . . . various municipal and public projects in the country—municipal water works, light, power and gas plants . . .—and also to conserve and utilize our natural resources of gas, oil, hydro-electric power."[32] Local representatives of this and other national and state urban reform groups consistently participated in East Bay public water campaigns. They helped define the agenda and goals of public waterworks. In 1918, however, their ambitious goals for public ownership frightened off as many voters as they attracted.

The comprehensive nature of the 1918 proposal aroused firm opposition from Oakland mayor John L. Davie, despite Oakland's involvement in the discussions that produced it. Davie was still wary of participating in any regional scheme that might weaken his power in Oakland. He countered the regional utility plan with proposals for a municipal waterworks modeled on Los Angeles' water district and city public service commission.[33] He argued that "the public utility district idea [was] fundamentally wrong in that it [sought] to establish a separate political unit distinct from both the municipality and the county." Davie criticized regional water because it would add to government bureaucracy and remove administrative functions from public control. He also predicted increased administrative costs of new agencies and cautioned that a poorly managed regional system would be "a hindrance to . . . local progress."[34]

Davie's opposition to the 1918 regional water proposal reflected both his rivalry with the Progressives and his insight into the political implications of a regional water agency. Unlike the Progressives, who were eager to turn services and administration over to appointed experts, Davie believed that elected officials should control services.[35] Furthermore, Davie had built his reputation as the East Bay's most flamboyant opponent of utility monopolies. He had contributed to Oakland's efforts to wrest the waterfront away from the railroad in the 1890s, and, by starting his own transbay service, broke the railroad's ferry monopoly in 1892.[36] A Progressive-sponsored public waterworks, particularly one that subsumed Oakland into a regional district, threatened to undermine Davie's base of power as well as his reputation as the East Bay's champion in the battle to control monopolies.

East Bay voters soundly defeated the public utility district in August 1918 even though war-related industries had created an additional drain of 3,000,000 gallons a day from the East Bay's overtaxed supplies.[37] Continued fear that new institutions would overburden taxpayers and decrease municipal autonomy turned many voters against the proposal. San Francisco and its Hetch Hetchy system still cast the shadow of annexation across the issue, and East Bay voters would not approve regional waterworks until they were assured that their municipal independence from San Francisco was protected. In addition, the fact that a water source had still not been identified worked against the proposal. In the absence of a feasible source, voters regarded public water proposals as empty gestures that would merely expand government authority and raise taxes, rather than as legitimate efforts to improve public services. As long as this was the case, the well-documented antipathy of East Bay residents towards the utility companies would not overcome their resistance to government growth.

East Bay Municipal Utility District

In the aftermath of the defeat of the 1918 Public Utilities District referendum, a new water advocate rose to prominence: Louis Bartlett, Berkeley's mayor and a well-connected East Bay Progressive. Bartlett believed that his community's approval of public water in 1914 and 1918 gave him a clear mandate and a solid political foundation from which to pursue regionalism. Furthermore, the East Bay's water crisis provided Bartlett with an exceptional opportunity to implement Progressive ideals of public ownership, good government, and scientific management. Bartlett was particularly well-suited to the water challenge. By soliciting active public participation in the planning process, he was able to create the broad coalition essential to a successful water campaign. He also managed to keep East Bay leaders behind the proposal even when they seemed most inclined to withdraw from the regional effort.

Many East Bay champions of public ownership, including George C. Pardee and John L. Davie, were accomplished politicians and leaders. Bartlett's political acumen alone cannot explain his success in winning approval of the East Bay Municipal Utility District. The political weight of decades of public water campaigns and the gradual erosion of utility power, signaled by expanded regulation of railroad, water, and other utility companies, came to Bartlett's assistance. So did the deplorable state of the East Bay water supply itself. A drought struck Northern California at the inception of Bartlett's campaign and lasted through the public referendum on EBMUD in 1923.[38] By 1920, overuse had reduced water levels to seven feet below sea level in fully eighty percent of East Bay wells. Four years later, water in these wells had dropped an additional eight to thirteen feet, making this source of supply extremely vulnerable to saltwater intrusion.[39]

Bartlett began his public water campaign as soon as he took office in 1919. Early that year, a San Francisco lawyer, E. S. Pillsbury, arranged an excursion for East Bay mayors to promote the Russian River as a water supply for their cities.[40] Bartlett took advantage of the renewed interest in public water development and convened an East Bay Water Commission made up of elected officials from Alameda and Contra Costa cities. Acknowledging Oakland's preeminence in the East Bay, Bartlett invited Davie to chair the Commission. Davie, however, remained wary of regionalism and soon stopped attending commission meetings. Bartlett saw Davie's withdrawal as an opportunity, a "free hand and . . . chance to do something."[41] Although not an official government body, over the next four years the commission hired consultants to study water problems, drafted legislation, and planned the political campaign necessary to manage a successful transition from private to public water supply. Commission meetings provided a forum for discussion of interlocal water options and regional administrative format and forged a spirit of cooperation vital to the success of Bartlett's project.

In 1919, the commission hired Philip Harroun to study the East Bay water situation. In the meantime, a new increase in water rates sparked renewed public enthusiasm for the commission's mission.[42] In May 1920, Harroun issued a report recommending that the East Bay build a regional waterworks using the Eel River as a source of supply. As had earlier studies, his report concluded that private water development interfered with long term planning and investment. To the disappointment of Bartlett, a longtime advocate of transbay cooperation, Harroun dismissed participation in San Francisco's Hetch Hetchy plan out of hand. He argued that a transbay system "would constitute . . . an economic mistake of the greatest magnitude, and should only be considered in case other sources . . . could not be secured." The economic risks lay not so much in the political arrangement between the two communities, he said, but in the high cost of the Hetch Hetchy project itself.[43]

Despite Harroun's warnings about Hetch Hetchy, Bartlett welcomed Harroun's conclusions that favored regional rather than municipal approaches to public water and asserted that only a cooperative system would give small cities access to the water they so desperately needed.[44] Nevertheless, the East Bay Water Commission hesitated. Most members of the commission believed that California law did not permit the level of interlocal cooperation that Bartlett and Harroun envisioned. Legislation passed in 1911 and 1915 did permit communities within a single county to join special districts, but the East Bay Water Commission hoped to establish a district that crossed county lines and encompassed the entire area served by People's Water Company's even larger successor, the East Bay Water Company. State law made no provision for this sort of regional cooperation.

Bartlett was not discouraged, however. In 1920, shortly after Harroun issued his report, he and the Alameda, Berkeley, and Oakland city attorneys began drafting new water district legislation. In early 1921, Bartlett turned over the draft bill to Alameda County legislators Anna L. Saylor and Homer R. Spence. Saylor had based her political career on limiting corporate profits and regulating the railroads and other private utilities. Spence went on to sit on the Alameda County Superior Court and the District Court of Appeals. Spence had been present at East Bay Water Commission meetings during the drafting of the Municipal Utility District bill.[45] Together, Spence and Saylor shepherded the Municipal Utility District bill through the Committee on Municipal Corporations and onto the floor of the California Assembly, where it passed unanimously in May 1921.

The Municipal Utility District Act granted regional utility districts significant municipal powers without actually interfering with city governments. The measure barred a regional district from acquiring "undesirable properties" or investing in useless projects, a clear allusion to the arguments that felled the Richmond Water Commission.[46] It also specifically prohibited cooperation between the East Bay and San Francisco, a provision that frustrated Bartlett early on.[47]

Under the act, the new district had to be approved by voters in each member town and to win support from city and then county governments in half the affected communities.[48] These provisions balanced the Progressive impulse to isolate critical planning and resource allocation functions from existing political structures with the existing distribution of power under electoral and local administrative regimes. As in Boston, the district at once preserved the autonomy of local governments and reduced their authority over water supply. Of course, in the East Bay, where the water company was suspected of corrupting local politics, this redistribution of authority was heralded as reinforcing rather than diffusing the power of voters and their representatives.

After the passage of the Municipal Utility District Act of 1921, the regional water campaign began in earnest. Bartlett and his allies did not want to see a repeat of the 1914 and 1918 water initiative debacles. So, before they approached voters with the referendum required by the Municipal Utility District Act, they assembled a broad coalition in favor of regional water.

A wide variety of civic organizations spoke out in favor of the proposed water district. For example, neighborhood improvement organizations and local Chambers of Commerce heralded the waterworks as a way to secure economic and political prominence for the East Bay. The California Federation of Women's Clubs looked forward to EBMUD promoting the principles of state forest and water planning. Bartlett encouraged such groups by inviting them to send representatives to the East Bay Water Commission meetings.[49]

To keep the public water question before voters, Bartlett made skillful use of the media. For example, he actively sought William Randolph Hearst's endorsement of the project because of Hearst's political influence. Hearst, of course, also wanted to sell newspapers, and had already used Oakland's water battles to boost circulation of his newly acquired *Oakland Post-Enquirer.* In a meeting with Hearst, arranged by Charles Summers Young, the *Post-Enquirer* publisher and Bartlett's personal friend, Bartlett convinced Hearst to use the paper to support regional water reform for the East Bay.[50] This was an unexpected victory because Hearst at the time wanted to offer California voters a reform alternative to Progressives like Bartlett. With Hearst's endorsement, all the major forces of reform were now joined together on the side of public water in the East Bay.

Because they believed that a regional water system would bolster economic growth, industrial leaders from Oakland, Berkeley, Emeryville, and Richmond also voiced their support and recommended that their municipal representatives pursue Bartlett's plan.[51] These entrepreneurs were well aware of the importance of water to their continued prosperity. The small water mains that ran to their plants did not deliver enough water for continued industrial expansion and could not provide adequate fire protection for the facilities that already existed. Some investors may also have realized that a regional water system would provide inexpensive public water service to unincorporated, unzoned, and other low-tax lands, making them attractive for industrial development.[52]

East Bay labor unions, too, joined the campaign for EBMUD, despite their differences with both industrial management and the Progressives.[53] The unions recognized that continued water problems would dampen economic growth and thus create hardship for workers. Many workers also hoped to secure jobs on the public construction projects. So, despite earlier suspicions of public ownership, the promise of jobs persuaded unions to join the EBMUD coalition.

Union endorsement of EBMUD was a sign of how effectively Hiram Johnson's opposition to railroads and support for pro-labor legislation changed political alliances in early-twentieth-century California.[54] Beginning with the 1906 prosecution of Abe Ruef, leader of the San Francisco Union Labor Party, labor had been wary of the Progressives.[55] Furthermore, because they had not organized public employees as effectively as private workers and because Progressive leaders' calls for public ownership frequently followed major strikes, unions saw the promotion of public utilities as a means to weaken labor.[56] At the same time, union members had no great affection for railroads or other private utilities. These corporations' history of hiring Chinese laborers in particular fueled nativism and resentment among California's labor organizations.[57] As was the case with so many groups in the East Bay, the unions' antipathy towards private utilities eventually won out over partisan rivalry.

Despite growing momentum, regional water was not yet assured of victory. Mayor Davie, still deeply suspicious of the regional effort, sought to derail Bartlett's plan with alternative proposals and by refusing to cooperate during legislative hearings on the proposed bill. He continued to warn that regional cooperation would weaken municipal autonomy. When the regional water district came before the public in 1923, Davie delayed announcing the election, refused to campaign for regional water, and insisted that EBMUD would have to be approved in Oakland without his endorsement.[58]

Davie's opposition was a serious obstacle. Oakland was the largest and richest of the East Bay cities, and no regional venture could succeed without it. A staunch believer in home rule, Davie dismissed regional projects as "extravagant, ponderous and inefficient" arrangements that could erode city self-government.[59] The East Bay Progressives' early enthusiasm for the Hetch Hetchy aqueduct confirmed, in Davie's mind, the connection between regionalism and the annexation of Oakland to San Francisco. This connection, fostered by Bartlett's initial insistence that the East Bay Water Commission not rule out Hetch Hetchy participation, steadied Davie's resolve. For all his opposition to regionalism, however, Davie did not advocate continued private ownership of local utilities. Rather, he favored municipal utilities as an ideal way to reinforce local independence and restore his own fading prominence.[60]

From the earliest discussions of regionalism, Davie had promoted a number of water schemes intended to reinforce Oakland's influence. In 1916, for example, he endorsed city-county consolidation. Such a comprehensive reorganization of local government could have allowed Davie to centralize the administration of a wide variety of services and build the kind of ambitious waterworks that the East Bay so desperately needed.[61] Davie even insinuated that his plan could allow Oakland to annex Berkeley and other communities. Consolidated administration of Alameda County, Davie insisted, would save public funds by lowering taxes, eliminating administrative redundancy, and closing the "open . . . door to the spoils system."[62] It is interesting to note the similarities between Davie's rhetoric and that used by Progressives during the same period. Both promoted the restructuring of municipal government as a means to reduce inefficiency, foster apolitical administration, and battle corruption. Neither Davie nor his Progressive opponents were disingenuous. Rather, both had fully accepted the proposition that political rivalry could and did interfere with public administration, and that private utilities, favoritism in hiring, and fragmented administration all hindered local government. As rivals, they simply disagreed about who should engineer the solutions to these problems and who should secure control over the new institutions created to solve them.

In the 1920s, Davie tried a different tactic. Faced with the impending referendum on the Municipal Utility District, he countered regionalism with suggestions that Oakland could easily develop local water sources for an independent municipal waterworks. Considering the decades of dire projections of local water shortages, Davie's proposal rang hollow and did little to dampen mounting public enthusiasm for a larger scale project.[63]

Davie's opposition to regionalism clearly discouraged other East Bay leaders, however. Almost immediately after the California Legislature passed the Municipal Utility District Act, most members of the East Bay Water Commission resigned. Bartlett, because he still insisted that the East Bay consider collaboration with San Francisco, bore a large measure of responsibility for this defection. Nonetheless, when the regional project appeared in jeopardy, he moved quickly, proposing that Berkeley and Albany build a joint municipal project. The East Bay cities fell for Bartlett's red herring and revived the Water Commission in order to defeat the plan and pursue regionalism. The *Berkeley Gazette* credited Bartlett's Berkeley-Albany plan with encouraging even Oakland to rejoin the movement. An Oakland newspaper celebrated the defeat of the Berkeley-Albany plan, touting truly regional approaches as the best means to secure improved water.[64]

The commission was resurrected none too soon. As Bartlett and his allies were moving the East Bay closer to regionalism, the East Bay Water Company stepped up its opposition to public enterprise, funding opposition groups, proposing new water projects, and lobbying hard. Wigginton Creed, the water company's president, later the head of powerful Pacific Gas and Electric Company, had built his career working for California's private water utilities. He held fast to the credo that private companies could provide water and other services more efficiently than could government agencies. "Efficient" in this context meant that the projects involved the smallest tax increases and least bureaucracy possible. This definition differed significantly from the Progressives' interpretation of the term, but attracted support from those who wanted both extensive services and low taxes. Like his fellow water, power, and railroad executives, Creed had fought a rising tide of government regulation ever since the Progressives' ascendancy in the 1910s. The passage of the Municipal Utility District Act in 1921 made his position even more precarious.

Creed used many strategies to try to prevent government takeover of East Bay water supplies. In June 1920, when the Water Commission began surveying the properties of the East Bay Water Company, in a first step towards purchasing the company's reservoirs and distribution system, Creed did all he could to interfere. He used company resources to campaign against the district, and predicted dire economic repercussions if the cities raised taxes to pay for EBMUD.[65] He also sought out—and created—sympathetic organizations to advocate private utility ownership in the name of the public interest.

Creed relied on civic organizations much as Bartlett did. Taxpayers leagues in Oakland and Richmond, possibly funded by the water company, echoed Creed's prediction that new water taxes would precipitate economic collapse. Although most commercial and industrial organizations in the East Bay pledged their support for EBMUD, the Berkeley Chamber of Commerce and the Oakland Manufacturers Association echoed the fiscal arguments of the taxpayers leagues. Creed and other water company leaders, who belonged to these organizations, clearly influenced their statements in favor of the private water system.[66] By 1925, Creed was using these groups to introduce alternative water plans to the East Bay public in an effort to undermine faith in the regional district. For example, the Oakland Manufacturers Association proposed that the East Bay water district purchase water from the Sierra Blue Lakes Power Company rather than build an independent water supply. This idea appealed to some frugal East Bay residents. The Sierra Blue Lakes plan would have put the East Bay at the mercy of a new utility company. George C. Pardee, former California governor and author of Oakland's first public ownership campaign in 1890, struck back, excoriating the Oakland Manufacturers Association as "enemies of municipal ownership," with "pecuniary interests" in private ownership. He portrayed the water company's opposition to EBMUD as an "un-American attempt to defeat the will of the majority."[67]

Eventually, the water company tried to counter the district's promise of new water supplies with proposals to expand its private waterworks. Charles Gilman Hyde, former director of sanitation for People's Water Company, developed plans to filter water from the Sacramento River.[68] Even in the 1920s, the Sacramento carried a heavy load of sewage and agricultural runoff. Nevertheless, Hyde asserted that the river, properly treated, would yield water "generally more satisfactory and probably more wholesome" than a mountain reservoir. Furthermore, filtering the Sacramento would cost twenty million dollars less than anything the regional district proposed.[69] In the end, the California Railroad Commission rejected the East Bay Water Company's permit application for the Sacramento project.[70] But the filtration plan appalled regional advocates. They accused the water company of demonstrating, once again, its lack of regard for the public's will and welfare.[71]

In the end, the voters in Alameda, Berkeley, El Cerrito, Emeryville, Oakland, and San Leandro demonstrated just how ineffective Creed's strategy was. In 1923, they voted two to one in favor of the East Bay Municipal Utility District. In 1925, they soundly endorsed EBMUD's construction bonds.[72] Throughout the campaigns, economic issues received the most attention. The promise of growth overcame public resistance to new bureaucratic institutions, large bonded debt, and the abandonment of California's long tradition of entrepreneurial development of public services. A major fire in the Berkeley hills in September 1923, a few months after voters approved the district, reinforced the impression that private water could not meet East Bay needs and, therefore, increased support for EBMUD.[73] Rivalry with San Francisco, and partisan competition between Democratic and Progressive city administrations also contributed to EBMUD's success.

The battle over regional water in the East Bay reflected long-standing competition among three dramatically different visions of California's future. The private utilities wished to keep politics out of resource allocation by keeping water and other development decisions outside the regulatory and administrative arena. In contrast, both the Progressives and Mayor Davie had largely adopted the Populists' late-nineteenth-century demands for public ownership of utilities. But Davie focused on the city as the source of influence and locus of political action, while the Progressives used water issues as a means to implement more comprehensive administrative reforms. Although Mayor Davie had a firm grip on Oakland's politics, his advocacy of local control did not prevail. Had EBMUD more substantially altered existing municipal and county governments, it would have encountered stronger opposition. Meanwhile, the Populist implications of public services attracted the large number of voters disaffected by the influence of the railroad and other private interests in state and local politics. In the East Bay, this antimonopoly sentiment translated into political attacks on the private water companies and ultimately led voters to accept the regional water system and the Progressives' political reforms.

Agriculture versus Industry

In 1924, the newly formed East Bay Municipal Utility District hired three engineers to design its public waterworks. That these engineers would seek water at some distance from the East Bay was, by this time, a foregone conclusion. They approached their charge technically, considering and discarding specific proposals primarily for reasons of feasibility, expense, and quantity of available water. They considered the potential costs of acquiring necessary water rights but did not discuss the effects that diverting water would have on communities already using the streams in question. By separating the selection of a water source from questions of water policy, EBMUD downplayed the resource conflicts inherent in the project. EBMUD, like Boston's Metropolitan Water Board, was prepared to transfer water from rural to urban uses and justify that transfer on economic grounds. This approach pitted farmers and irrigation developers in rural communities against California's third largest urban area. The conflict was more than just a battle over water rights; it was part of a series of protracted debates over economic development, public administration of natural resources, and the balance of political power in the state.

To design the new water system, the East Bay Municipal Utility District hired Arthur Powell Davis, William Mulholland, and George W. Goethals, three of the most prominent engineers in the nation. William Mulholland was best known as the architect of Los Angeles' municipal waterworks and the Owens Valley Aqueduct. Davis, former head of the United States Reclamation Service, had opposed Mulholland's plans for the Owens River. Goethals had made his reputation as the builder of the Panama Canal. Faced with a large territory and urged to speed their investigations, Davis, Mulholland, and Goethals consulted with local experts. They sent questionnaires to companies and individuals with "water and power projects for sale" requesting information about water volume, type of water rights held, size of watershed, power potential, and types of projects considered for the site, as well as the price of the whole package.[74] These questionnaires and a brief study of earlier water reports yielded eight possible source rivers: American, Eel, Feather, McCloud, Mokelumne, Sacramento, Stanislaus, and Tuolumne.

The engineers included the Tuolumne because Bartlett and a few other EBMUD directors were still promoting the Hetch Hetchy project, despite provisions in the Municipal Utilities Act that excluded this option.

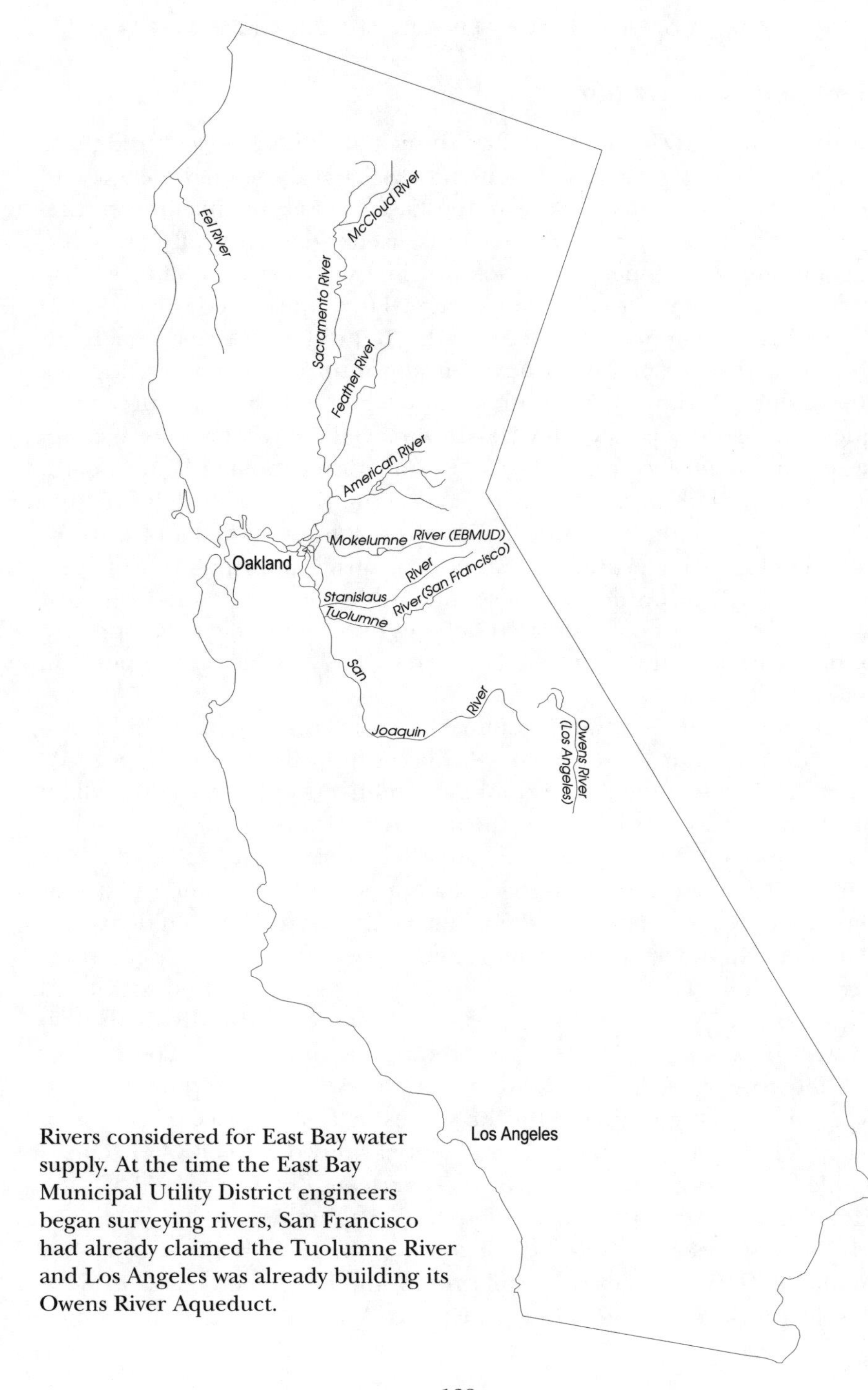

Rivers considered for East Bay water supply. At the time the East Bay Municipal Utility District engineers began surveying rivers, San Francisco had already claimed the Tuolumne River and Los Angeles was already building its Owens River Aqueduct.

But this river proved one of the easiest to eliminate. By this time, Hetch Hetchy had become the "political football of San Francisco politics" and would not be operational for many years. Moreover, San Francisco would not guarantee the East Bay unimpeded access to water.[75] As late as June 1924, San Francisco's Mayor Rolph refused to negotiate with EBMUD before "the East Bay directors had satisfied themselves as to San Francisco's title to the water."[76] He implied that the East Bay cities would have no rights to the Tuolumne, an arrangement that would have made them dependent on San Francisco. Bartlett interpreted Rolph's approach as an attempt to force the East Bay to accept annexation to San Francisco.[77] Meanwhile, the Hetch Hetchy League of the East Bay Cities stridently criticized the engineers' decision to reject the Tuolumne and threatened to become the nucleus of East Bay opposition to the new water district.[78]

Of the seven remaining rivers, Davis and his fellow engineers judged the Mokelumne, a tributary of the San Joaquin River located a few basins north of the Hetch Hetchy Valley, to be the most promising. Developing the American would have required the district to build many costly small reservoirs. The city of Alameda already owned rights on the Eel and offered to transfer these rights to EBMUD free of charge. However, Marston Campbell, EBMUD's first president, argued that the Eel River project played into the hands of the East Bay Water Company and the anti-EBMUD Oakland and Alameda administrations; moreover, Davis judged the river to be too small.[79] The Feather already supported many irrigation and power projects. To develop a water project there, EBMUD would have had to spend huge sums condemning land and water rights. The McCloud and the Stanislaus were each too small to supply the region. The Sacramento was simply too filthy.[80]

The Mokelumne had seen considerable development by the time the East Bay engineers began their survey. Hydraulic mining had scarred the watershed, forcing salmon from the river by 1856.[81] Pacific Gas and Electric and other utilities had built hydroelectric generating dams in the Mokelumne's high mountain canyons. Below, in the flat, fertile Sacramento–San Joaquin Delta, pumps and canals carried water from the Mokelumne into the vineyards and fields that spread out on either side of the river. The Lancha Plana basin, located in the Sierra foothills and identified by Davis as the best site for EBMUD's Mokelumne reservoir, had attracted the attention of two independent water developers, Charles Landis and Stephen Kieffer. They had spent several years attempting to build an irrigation and electric-generating project at Lancha Plana. Kieffer had bought out his partner and onetime employer by the 1920s but had been unable to attract the investors he needed.[82] He brought his project to Davis's attention before the EBMUD engineers began their survey, hoping that he could recover his investment by selling the project to the East Bay water district.

Kieffer's plan appealed to Davis, Mulholland, and Goethals. The Mokelumne still contained some unappropriated water, few people lived in its watershed, and the aqueduct route could incorporate an emergency water intake from the San Joaquin River. Perhaps most importantly, the Lancha Plana basin could hold much more water than the other projects under consideration, particularly if supplemented by the neighboring Arroyo Seco basin.[83] In 1925, following the recommendation of Davis, Mulholland, and Goethals, EBMUD applied to the Division of Water Rights for a permit to dam the Mokelumne River at Lancha Plana.

In September 1925, the Division of Water Rights held hearings on the Mokelumne water rights. The agency had two other permit applications on file for the Lancha Plana dam site. Kieffer had submitted one. The other had come from a consortium of prominent San Francisco investors headed by John W. Preston, Jr. The hearings quickly became a forum for debate not only on the relative merits of irrigation and urban water use, but also on the state's economic priorities. Each side attempted to use California's water distribution statutes to its own best advantage. EBMUD asserted that, as a municipal agency dedicated to providing domestic water supplies, it had priority over all other claimants.[84] Meanwhile, Kieffer, Preston, and other representatives of San Joaquin valley agriculture argued that state law protected irrigation.

Preston and Kieffer represented fundamentally different approaches to private water development. Preston applied to build a multiple-use irrigation and hydroelectric power project. But his proposed irrigation works were largely a front for the more lucrative hydroelectric power development since California law gave agricultural use priority over hydroelectric generation.[85] When, in October 1925, Preston announced that he had signed contracts with Pacific Gas and Electric Company for the power generated from his water rights on the Mokelumne and Yuba Rivers, he exposed his strategy and undermined his permit application.[86] Although utilities had frequently used this precise strategy to claim water resources, a power and irrigation project could no longer compete with domestic waterworks in the eyes of the Division of Water Rights.

Kieffer, who by 1924 owned substantial holdings in Lancha Plana—including the proposed dam site—did not want to compete with EBMUD as much as he wanted to profit from it. His interest lay in winning construction contracts or securing generous compensation for his lands. He applied for a permit to develop Lancha Plana in order to increase both the value of his property and the amount of compensation for his lands that he could demand from EBMUD. That he offered to surrender his property in exchange for a contract to build the whole Mokelumne system and then sued for larger compensation payments after EBMUD condemned his holdings exposed him as a profiteer.[87]

During hearings to settle the compensation payments, EBMUD discovered that Kieffer had purchased a significant portion of his Lancha Plana holdings after EBMUD engineers saw his plans for the Mokelumne and after EBMUD adopted those plans in 1925. Armed with this information, the district convinced a judge that EBMUD should repay Kieffer only the three hundred thousand dollars he had actually spent on his Lancha Plana lands.[88] Kieffer did not abandon speculation with Lancha Plana; he would go on to seek outrageous compensation for a reservoir site on the American River and a road right-of-way needed in Calaveras County.[89] This sort of speculation bedeviled much public utility development in California; Davis, Mulholland, and Goethals, by requesting project suggestions from their owners, only encouraged such opportunism. However, because Kieffer and his ilk were widely seen as reaping unearned profits and interfering with efficient growth and with the implementation of important public policies, they inadvertently fostered support for government water planning.

In 1925, the Division of Water Rights granted EBMUD's permits on the grounds that domestic water use took priority over irrigation and hydroelectric generation, and that public projects superseded private development. Kieffer and Preston sued. They challenged the decision on the grounds that they had filed for Mokelumne water before the District had. Kieffer went further, charging the District with copying his original Lancha Plana application and then using its influence to delay processing of his permit.[90] Both Kieffer and Preston insisted, however, that their dispute with the East Bay did not merely turn on questions of fairness. They argued that the state water board should protect private rather than public enterprise, and favor rural rather than urban water needs.[91]

Preston's lawyers stated the case for irrigation and against EBMUD most cogently. They pointed out that EBMUD's municipal system would carry water for nonhousehold activities that did not fall under the "domestic" category and therefore should not supersede Preston's application to use the Mokelumne for irrigation. Street sprinkling, sewer flushing, and water used in public buildings or for fire fighting were not "a domestic use" and did not have any connection with the home as such." Furthermore, industry was no more domestic than sewer flushing. To grant water to the East Bay for industrial development, they complained, would sacrifice the agricultural interior for the growth of coastal industrial cities.[92] This argument resonated with many farmers in the Mokelumne Valley. By blending a defense of private ownership with anti-urban complaints, Preston, and later the city of Lodi and other opponents to EBMUD, offered far more effective resistance to metropolitan claims than did their Boston counterparts.

Preston and Kieffer were not the only ones to speak out against the East Bay Municipal Utility District's Mokelumne plans. Local newspapers—the *Stockton Independent,* the *Stockton Record,* and, especially, the *Lodi Sentinel*—also inveighed against urban development of water that San Joaquin Valley farmers needed. The Woodbridge Irrigation District, the Calaveras Water Users' Association, and representatives from the potato-growing community on nearby Staten Island also attacked EBMUD. They claimed that they and other users had already appropriated all the water in the Mokelumne. The Lawrence Holding Company and the Delta Land Syndicate focused on future rather then existing water needs in the San Joaquin Delta and near the Mokelumne.[93] They insisted that diversion of this water would prevent future agricultural expansion. They argued that because "in the north Pacific coast territory . . . fully NINETY-FIVE percent of all available water supplies can NEVER be used for agricultural or domestic purposes" because so few people lived there, the East Bay should take water from northern rivers and leave southern waters for the farmers.[94]

Civic organizations, including the Calaveras, Amador, and Tuolumne County Chambers of Commerce, proposed legislation that would have entirely prohibited Bay Area cities from taking water from San Joaquin River tributaries.[95] A more practical proposal called on state water officials to reserve a small percentage of the water originating in any given county for future use in that county.[96] Supporters defended this proposal on the grounds that water policies should promote balanced economic development throughout California, and that local irrigators had a greater right to their rivers than did distant communities. At the time, however,

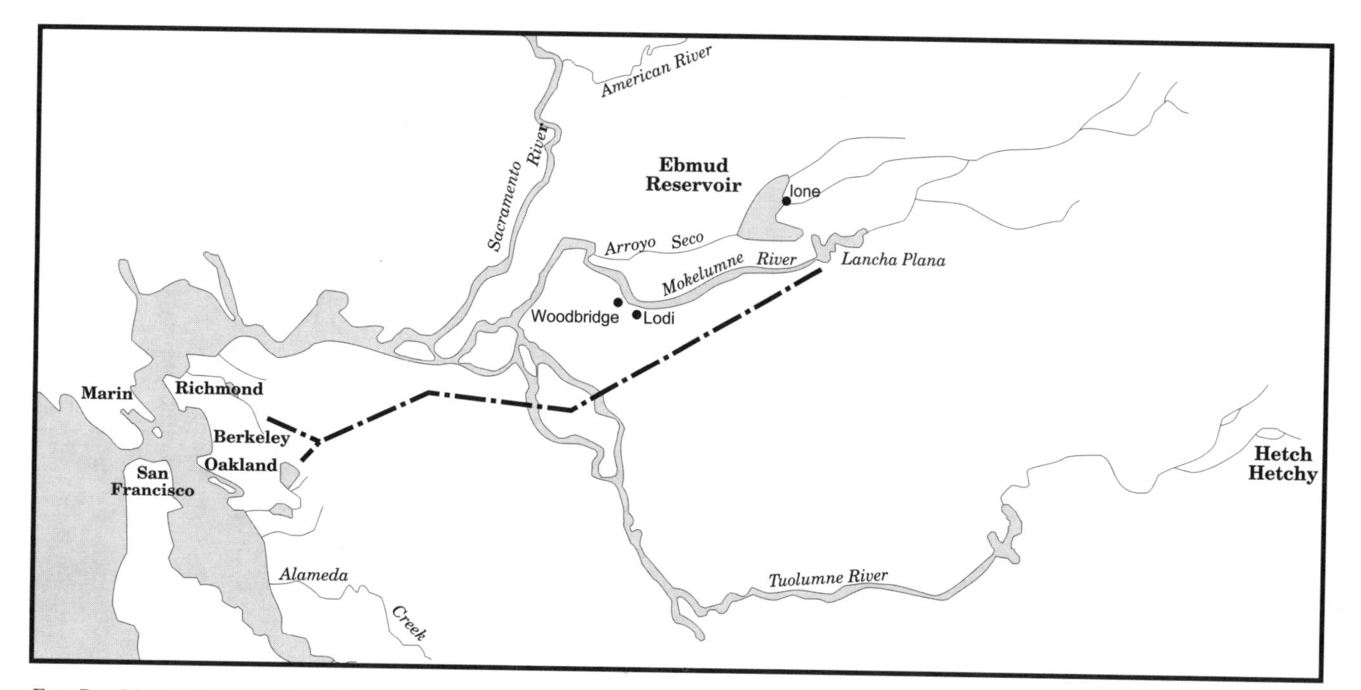

East Bay Municipal Utility District water system. The Mokelumne River reservoirs in Lancha Plana and Arroyo Seco angered farmers living near Lodi and Woodbridge because they feared that EBMUD's system would interfere with irrigation.

the State Division of Water Rights distributed water rights only to individuals and institutions seeking permission to pursue specific projects. This proposal, which might have secured water for future growth in rural areas, nevertheless flew in the face of too many of California's water rules.

After the hearings, in November 1925, the Lodi Chamber of Commerce and 100 area farmers proposed a San Joaquin County irrigation district. The Chamber of Commerce promoted the new district as a means to establish new water policies for the region and to "deal with the present Mokelumne river situation either through an agreement with . . . [EBMUD] or through the courts in an effort to protect existing rights."[97] The Chamber hoped that an irrigation district would organize opposition and give Lodi farmers the kind of institutional advantage that EBMUD enjoyed. The Lodi group faced considerable opposition, however. Many of the area's largest landholders, including the Southern Pacific and Western Pacific Railroads, petitioned to exclude their lands from the proposed district because they had ample irrigation water or did not want to pay for the drawn-out litigation that might result from conflicts between the district and EBMUD.[98] Because the irrigation district proposal identified no water source, some opponents feared that they would have to purchase water from Preston or other "soulless" utilities.[99] This last resembled the anti-utility sentiment that had driven the East Bay's public water campaign. In view of EBMUD's legal advantages, it is not clear that an irrigation district could have effectively opposed the urban water project. In any event, the irrigation district was defeated, 911 to 467, in a May 1926 special election.

By 1926, Stockton had contracted with EBMUD to store its own municipal water supply behind the East Bay's dam.[100] This agreement reversed that city's earlier opposition to the Lancha Plana reservoir. Stockton's decision and the defeat of the irrigation district left the City of Lodi alone in opposition to EBMUD. As administrative hearings and institutional remedies had failed, Lodi took legal action to halt East Bay construction. First, it challenged EBMUD's plans on the grounds that existing claims accounted for all the water in the Mokelumne. Later, Lodi challenged EBMUD's power to condemn city lands located along the aqueduct route. In additional cases, farmers argued that interbasin transfer would reduce groundwater recharge.[101] While important to Lodi and the East Bay and effective in highlighting water policy issues that had emerged in both the water hearings and public discussions of the Mokelumne project, these cases established no crucial legal precedents.[102]

Of all of the decisions in the Lodi cases, the one addressing the important water rights issue was handed down by the Court of Appeals in 1936, seven years after Mokelumne water reached the East Bay. Based on evidence that the Mokelumne provided the bulk of the water in Lodi's wells, a lower court had sought to protect Lodi's groundwater supply by ordering EBMUD to maintain minimum water levels in the Mokelumne River by releasing water from the Lancha Plana reservoir. The Court of Appeals overturned the lower court's decision on the grounds that only a small amount of water found its way from the river to the aquifer, while the rest flowed unused to the sea. Wasting a large amount of water to protect the wells, particularly when so much of the groundwater overdraft was caused by irrigation, did not meet the "reasonable use" standard established by 1928 law. The Court of Appeals ruled that if EBMUD's project did endanger Lodi's water supply, the district must provide Lodi with water or increase the water level in Lodi's wells.[103] This decision, together with earlier court defeats, effectively deprived Lodi and its surrounding irrigators of their claims on the river and guaranteed that future Mokelumne water development would primarily benefit East Bay industry.

With its bonds and water claims approved, in September 1925 EBMUD began constructing the Mokelumne dam and aqueduct. By the time the project was completed in June 1929, East Bay reservoirs "were all but empty" and wells were "on the verge of becoming saline," but the East Bay's future was secure.[104] The East Bay Municipal Utility District had succeeded where similar campaigns failed because the public was convinced of the clear financial and public service benefits regionalism offered. In this, the EBMUD campaign closely resembled its Boston antecedents. The two regions, however, built their networks in response to different environmental, political, and social conditions. Boston embraced regionalism in response to water quality problems, while the East Bay proved more concerned with water quantity. The differences reflect distinct historical periods, geography, and the nature of municipal authority, but should not overshadow the similar patterns of regionalism that developed in both of these communities.

Regional Sewerage: EBMUD's Special District One, 1937–1944

From 1929, East Bay residents could enjoy water from the Pardee Reservoir on the Mokelumne River. Regional sewerage—and, in fact, any significant changes in East Bay waste disposal practices—however, remained more elusive. As new highways and bridges routed more traffic along the waterfront, sewage pollution attracted more attention and, as a result, more protest. By the time EBMUD began water service, seventy years of haphazard sewage disposal had fouled the shorelines of Oakland, Emeryville, Berkeley, and Richmond with organic and inorganic wastes whose stench filled the air. More than 60 municipal sewer outfalls were spewing raw sewage onto tidal flats and inlets all along the East Bay waterfront.[105] Engineers studying East Bay sewerage found "quite uniform evidence of sewage pollution" including "floating or stranded fecal and other sewage particles"[106] According to one report: "the continuous deposition of sewage [on tidal flats] . . . formed a black organic putrescible mud that covers much of the area. On warm days and with suitable wind conditions, the odors resulting from the decomposition of this organic matter can be identified at a distance of two miles or more inland."[107]

Much of the pollution was caused by the Standard Oil refinery and the food canneries that lined the waterfront from Richmond to San Leandro. Surveys of the East Bay revealed waterways littered with vegetable remains and stained with liquids from fruit processing; small creeks ran dry but for the waste water running through them.[108] Canneries disposed of large quantities of vegetable and fruit wastes faster than marine organisms could digest them. The bacteria that assisted decomposition of organic wastes used up dissolved oxygen along the East Bay shore before it could be replenished by wave action and tidal currents. One study of the sewers revealed that dissolved oxygen levels around sewer outfalls and inlets had fallen by fifty to one hundred percent to levels so low that no significant fish or marine life could survive.[109] In the absence of adequate oxygen, anaerobic bacteria replaced the oxygen-dependent species, releasing the foul-smelling gasses that so offended travelers along the waterfront. One compound in particular, hydrogen sulfide, caused "an inestimable amount of damage to the paint of buildings and marine structures" along the harbor and to houses in eastern Alameda. Pollution from domestic and industrial sources had rendered the East Bay waterfront foul and lifeless.

Despite the alarming reports, the impact of sewage pollution on marine life was of minor concern compared to its effect on industrial development. Charles Gilman Hyde, an engineer who studied East Bay sewerage in 1937 and again in 1941, labeled sewage pollution "a handicap to industrial development and shipping."[110] Citizen groups such as the

Pardee Dam on the Mokelumne River, completed in 1929, after just four years of construction. Courtesy East Bay Municipal Utility District, Oakland, California.

Lancha Plana dam site, 1927. Courtesy East Bay Municipal Utility District, Oakland, California.

East Bay contractors had more machinery at their disposal than did their Massachusetts counterparts. Courtesy East Bay Municipal Utility District, Oakland, California.

Western Water Front Industries Association, the West Oakland Booster's Club, and the Waterside-Thompson Improvement Club agreed that the East Bay's lack of adequate waste disposal facilities interfered with local growth and development.[111] By the 1920s, when these organizations began their campaign to improve shoreline conditions, advances in medical science had all but eliminated the association between decay and disease. Germs had replaced miasma as the critical health concern. And because few people had contact with East Bay water, boosters and engineers alike paid little attention to potential health risks posed by the existing or proposed sewers. Here, as during discussions of the public water supply, the salient concern was economic development, not disease and social chaos.

The germ theory did introduce one health-related element to discussions of regionalism—contamination of shellfish with infectious agents. Filter feeders like oysters and other bivalves concentrate wastes and pollutants, and can spread cholera, hepatitis, and other waterborne diseases. In 1916, even before bayshore odors attracted much attention, the California State Board of Health noted that sewage disposal was contaminating oysters in San Francisco Bay, damaging the profitable Bay oyster industry, and possibly endangering the health of people who ate Bay oysters. The board recommended an analysis of the situation to determine "whether the [oyster] industry [was] worth the cost of protecting the [oyster] beds from pollution."[112] One could not, the Board implied, have both oysters and inexpensive sewerage. With this statement, the Board of Health identified sewerage as fundamentally a question of resource distribution. By the end of World War II, manufacturing would clearly win the battle between inexpensive waste disposal and healthy marine life. Even as early as 1916, East Bay leaders identified their economic future as dependent upon industrial growth. As a result, they clearly favored inexpensive sewage disposal, regardless of the consequences for the oyster industry.

In the late 1920s, drainage problems and damage caused by hydrogen sulfide gas finally broke through the inertia that had prevented

Cleaning bedrock and pouring concrete for Pardee Dam, 1928. Courtesy East Bay Municipal Utility District, Oakland, California.

Pardee reservoir spillway, 1929. Courtesy East Bay Municipal Utility District, Oakland, California.

action even in the wake of the ominous report about Bay shellfish. In 1927, storm runoff from the hill communities of Piedmont and North Oakland flooded flatland neighborhoods in Oakland, Berkeley, and Emeryville. These towns responded by authorizing a three-member Engineering Commission on Sanitation and Drainage to develop a cooperative plan to reduce runoff that crossed city boundaries. Drainage commissioners Charles D. Marx, Charles H. Lee, and Harold F. Gray concluded that the East Bay cities needed to coordinate both storm and sanitary sewerage as a single, united district. Within months, however, plans for regional storm drainage had bogged down in a morass of proposals and counterproposals. Engineering problems and the Depression delayed further progress until the late 1930s.[113] Nevertheless, subsequent

(Facing page, top) Calaveras Cement Company used their participation in East Bay Municipal Utility District dam construction for their advertising. Waterloo, California, 1929. Courtesy East Bay Municipal Utility District, Oakland, California.

(Facing page, bottom) Pardee dam contractors provided mess halls, a general store, and other facilities in the construction camp. Courtesy East Bay Municipal Utility District, Oakland, California.

3,000,000 Sacks
of
CALAVERAS CEMENT
Being used in Mokelumne Project
Pardee Dam

DO NOT PARK HERE

discussions thoroughly incorporated the engineers' diagnosis that the East Bay's drainage problems were regional.

Oakland City Engineer Walter N. Frickstad resurrected regional sewerage in a 1936 report on waterfront odors requested by the Oakland City Council. Frickstad echoed the 1927 Engineering Commission's recommendation that the East Bay construct a joint outfall into the deep waters of San Francisco Bay. He did not pursue this position long, however, scaling back his proposal to a simple cooperative project between Oakland and Emeryville. Meanwhile, the New Deal was creating new opportunities for public works, and Oakland leaders saw in Frickstad's report the chance to use federal funds to eliminate waterfront odors while also employing hundreds of the Depression's victims.[114] Engineers from nine East Bay cities rushed to prepare 1938 Public Works Administration project applications, but missed the deadline.[115]

The East Bay failed to meet the critical 1938 deadline because no single agency would accept responsibility for the project. Beginning in 1926, each proposal for regionalism had named EBMUD as the best agency to oversee a metropolitan sewerage system. However, district directors argued that they did not have authority to take on such a project. Not until the legislature amended the Municipal Utility District Act to permit the district to create special subdivisions for sewage disposal would EBMUD agree to participate.[116] In the meantime, the United States' entry into World War II delayed further action. Unfortunately, the rapid development of war industries in Richmond and elsewhere in Alameda and Contra Costa Counties heightened the need for improved sewerage in the region.[117]

The legislature gave EBMUD the power to create sewage subdivisions in 1941. As the war drew to a close, EBMUD sought voter approval of a sewerage division, Special District 1. Directors intended to have a plan ready to take advantage of postwar federal programs to reconstruct the civilian economy. The new special district was to include only Alameda, Berkeley, Emeryville, Oakland, and Piedmont, a fraction of the territory served by the EBMUD waterworks. With EBMUD finally empowered to build the kind of regional sewerage that the East Bay needed, and with ample federal funding available, Special District 1 faced only minimal opposition. A few critics saw the district as too small to reduce pollution or facilitate East Bay growth. Some pro-growth groups, such as the Oakland Real Estate Board, objected to the omission of the unincorporated areas, prime locations for industrial development. On the whole, however, industry and commercial interests took a stance in favor of public works as they had in the 1920s EBMUD battles. Labor, in contrast, split over Spe-

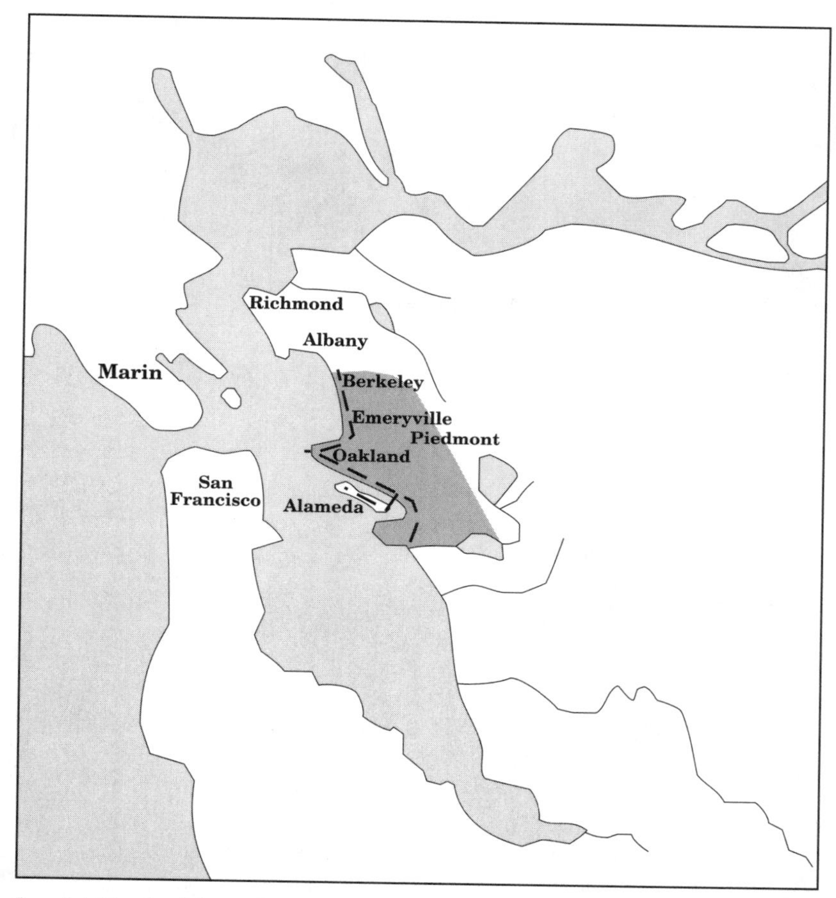

Special District 1 brought regional sewerage to the Alameda County portion of the East Bay Municipal Utility District after World War II.

cial District 1. The American Federation of Labor opposed the district on the grounds that it placed too much power in the hands of EBMUD directors, and that voters should retain more influence over the final sewer project. Meanwhile, the Congress of Industrial Organizations adopted the pro-growth stance articulated by boosters and manufacturers.[118]

Even though economic growth, not public health, provided the impetus behind Special District 1, the California Department of Public Health provided consistent and forceful support for regional sewerage. As early as 1937, the Board of Health had urged the East Bay to take action to

Settling tanks and flow meters for Special District 1, the East Bay's regional sewerage system. Courtesy East Bay Municipal Utility District, Oakland, California.

reduce shoreline pollution. In 1939, health officials despaired that "the trend has been more strongly toward abusing harbor waters in sewage disposal than it has been away from it. But the reasons . . . are not technical, so much as they are financial and political."[119] In 1944, the Board returned to the East Bay sewerage discussions, urging voters to approve Special District 1. As the state agency in charge of monitoring wastewater facilities, the Board of Health's support was critical but it reflected the delegation of water quality issues to health agencies, not the kind of activism seen among Massachusetts and Boston health officials in the nineteenth century. In 1944, voters did approve Special District 1, opening the way for regional sewerage.

In the postwar years, institutions like state boards of health and, later, environmental regulatory agencies, stepped in to promote public works and water quality much as political and social reformers had done in the earlier period. Because water supply and sewage disposal systems in the East Bay were separate and because the germ theory narrowed perceptions of the health hazards associated with water pollution, East Bay communities delayed sewerage improvements for far longer than Boston did. Despite these differences, regional sewerage in Boston and the East Bay shared some basic goals and consequences. Both systems narrowly addressed local health and economic needs, albeit the commercial implications of sewerage were more pronounced in the East Bay. By perpetuating inexpensive waste disposal, Special District 1 sacrificed Bay Area fisheries, essentially nonurban Bay resources, in favor of urban and industrial growth. In this way, the sewage system repeated the resource allocation decisions made throughout the history of urban regionalism. Rural activities and economies were sacrificed for urban prosperity, in spite of continued nostalgia for America's rural past.

Implications of East Bay Regionalism

By the time East Bay voters approved the East Bay Municipal Utility District, urban regionalism was well established in American politics. Few citizens disputed the idea that semiautonomous, specialized agencies could administer public services better than municipal officials could. Federal spending during the New Deal and postwar reconstruction programs further reinforced the trend towards regionalism because it solved the problems of financing large, expensive projects.

In the East Bay, the transfer of water supply and sewerage from private to public hands and from municipal to regional administration represented a triumph of the Progressive vision. The creation of EBMUD, charged with managing natural resources, promised to rationalize and stabilize economic development in the East Bay.[120] On the other hand, the existence of public agencies like EBMUD raised profound questions about the direction of California water policy. Some of the questions were familiar. Should private or public agencies control California's natural resources? What types of resource uses would best ensure growth and wealth? Others, however, were new. Could water laws that had evolved to regulate individual claims provide for the equitable distribution of resources among communities? Should cities' claims overshadow those of private water users? And most importantly, should cities be allowed to take water from distant rivers to ensure their future prosperity, even if such interbasin transfers significantly curtailed potential for development in the source regions?

As interbasin transfers like the Mokelumne project increased water competition throughout the state, many of the same advocates of public ownership who had backed EBMUD now endorsed state-sponsored irrigation development. They argued that centralized water administration at the state level would ensure continued agricultural prosperity and reduce sectional bitterness. Regionalism, they pointed out, had eliminated interlocal water competition in urban areas. Now they wanted to use similar organizations to settle resource conflicts throughout the state. These proposals, outgrowths of water resource competition and Progressive notions of scientific resource management, eventually yielded the Central Valley Project and State Water Plan.[121]

As with urban regionalism, the rising clamor for centralized irrigation planning reflected economic priorities. The California legislature had taken responsibility for establishing water use priorities early in the state's history and had continued to do so in an effort to manage California's economy. For example, in 1855, the California Supreme Court approved legislation intended "to foster and protect the mining interest as paramount to all others" in water rights.[122] Mining retained its protected status for as long as it dominated the state's economy. Toward the end of the nineteenth century, however, the growing importance of agriculture and fishing, and the recognition that they had greater long-term economic potential than mining, led the state to guarantee greater water allowances for farms and fisheries.[123]

The East Bay submitted its permit applications for the Mokelumne River during another shift in California's economic development and, consequently, during the transition to a new water policy. From 1900, industry began challenging agriculture's status as California's hope for prosperity. Even the state irrigation agency acknowledged that any centralized water plan had to ensure adequate supplies for industry.[124] Manufacturing appealed to California policymakers who believed that it had greater potential for growth and stability than did farming. Industry offered jobs for California's urban dwellers, jobs that did not fluctuate with drought, crop diseases, and commodity prices.

Gross receipts from East Bay plants demonstrated industry's growing importance in California. In 1920, the value of industrial products in Alameda and Contra Costa counties was estimated at just over $500 million, only $90 million less than the state's total agricultural output.[125] East Bay industry underwent a spectacular boom between 1910 and 1930; the number of manufacturers more than doubled, and the number of employees increased almost sixfold. This growth placed enormous pressure on state officials to make water supplies available.[126] Because manufacturing did not appear on the Division of Water Rights list of water use priorities, irrigators continued to dispute the legitimacy of water distributions made to industries. However, because industrial consumption was subsumed under urban water use, the economic importance of industry contributed to cities' ability to secure permission to develop new water supplies from the Division of Water Rights, regardless of established protections for California's agriculture.

Whether they wanted to cooperate or even merge with larger neighbors or use public waterworks to reinforce municipal autonomy, city leaders supported the transfer of water from rural to urban communities. Given the East Bay's chronic water shortages and growing difficulties with contamination, access to rural rivers was crucial to their cities' futures. Furthermore, proponents of interbasin transfers had historical precedent on their side. Cities had long extended their reach to new sources of water to meet growing domestic and industrial needs. Boston's Wachusett Reservoir, of course, but also New York's Croton Aqueduct and Los Angeles' Owens Valley showed California leaders that cities could justify taking distant waters.

Because they were following precedents, California's elected officials were largely unprepared for the opposition to their plans to transfer water. Los Angeles displaced agriculture in the Owens Valley so callously that farmers retaliated by bombing Los Angeles waterworks no fewer than eleven times. The first attack was in 1924, just eighteen months before the State Division of Water Rights convened hearings on the East Bay's Mokelumne plans.[127] The East Bay's project never elicited violent protest, in part because EBMUD recognized all municipal and agricultural water claims on the lower Mokelumne. Nevertheless, Lodi ranchers cited the Owens Valley controversy to inspire opposition to EBMUD, calling their community to arms with references to "the tragedy of Owens Valley . . . where hundreds of ranchers . . . took the law into their own hands upon their deprivation of life-giving water for their farms."[128] The Owens Valley bombings manifested urban-rural conflict at its worst. They highlighted the vital role that water played in California's economy and the weaknesses of the system designed to apportion water among potential users. In the aftermath of the Owens Valley episodes and protracted litigation over the EBMUD project, only state-wide water planning seemed likely to guarantee consistent economic growth in both rural and urban areas.

By the 1920s, farmers in the Owens and Mokelumne valleys had gained powerful allies. In several statewide elections, including the election of Governor Friend Richardson in 1922, voters rejected Progressive-style centralization and urban-focused policies. A onetime Progressive, Richardson now attacked his former allies' agenda. He shelved some of the social service commissions that earlier administrations had created and assisted legislative efforts to weaken a number of California's election reforms.[129] More significantly for farmers, Richardson opposed the transfer of water from rural to urban areas of the state. He forced the East Bay cities to test EBMUD construction bonds in the courts, declaring, "The water in the Mokelumne river is for the use of residents of the territory through which it flows. . . . I want to see legislation enacted that will save such rivers as the Mokelumne for the use of the local residents."[130] Richardson's determination to block EBMUD did not mean he was unwilling to use the powers of state government to direct economic growth and to develop natural resources. On the contrary, he championed an ambitious water conservation and redistribution plan intended to increase the amount of water for irrigation and promote "full development of the state's water resources" by diverting waters southward from the northern and central Sacramento and San Joaquin valleys.[131]

Richardson was not the first or only Californian to advocate state irrigation development or comprehensive water planning. In fact, Californians began clamoring for state or federal irrigation assistance in the 1890s. In 1919, Robert Bradford Marshall began to campaign for multiple-use development of the Sacramento watershed; subsequent projects implemented many aspects of the Marshall Plan.[132] From the 1930s through the 1960s, proposals for state irrigation development would be as ubiquitous in California politics as attacks on the railroad had been from the 1870s through the 1920s. Although Californians might disagree over how and where to build a state water system, few would speak against it. Louis Bartlett and George C. Pardee, the East Bay's most vocal public water advocates, agreed with Richardson in this one area. When he took office in 1918, Bartlett saw public ownership not only of the East Bay water supply, but also of water and timber resources throughout the state, as the key to efficient, sensible administration and economic development. He argued that corporate competition for lucrative municipal contracts corrupted local officials, and also that the activities of utility monopolies outside urban areas led to extraordinary waste of natural resources. Like the farmers hurt by interbasin transfer, Bartlett saw increased state action as the best way to ensure prosperity and ample resources for all.[133]

In Lodi, fears that EBMUD's project would severely restrict future agricultural expansion led to demands for coordinated water planning like that envisioned by Richardson and Bartlett. During the fight for the Mokelumne, Lodi area residents repeatedly called for intensive water development, including "the ultimate erection on every stream flowing from the Sierra Nevada Mountains . . . an impounding dam sufficient to conserve the waste waters . . . for irrigation."[134] Even opponents of EBMUD who acknowledged the coastal cities' water need, insisted that no "river should be given into the control of any one interest . . . [but] must be awarded to meet all needs."[135] In the long run, meeting all of California's water demands would prove an impossible task, but comprehensive water planning permitted greater consideration of future conditions than did the Division of Water Rights' limited power to choose among conflicting water project applications.

Despite its demands for state-sponsored water administration, Lodi's record on public water revealed an ambivalence toward increased government authority shared by many Californians. Voters supported public development that directly benefited them, but regarded government growth with suspicion when it less directly improved their lot. Thus, the *Pacific Rural Press* excoriated municipal utilities because they increased rural tax burdens by abolishing taxpaying private utilities.[136] And so, despite regular calls for coordinated resource development, California voters soundly rejected legislation that would have permitted the state to build waterworks for irrigation, the Water and Power Act, in 1922, 1924, and 1926. Even those who endorsed public ownership in general criticized plans for a state irrigation project as a "radical and serious" step that effective regulation had rendered largely unnecessary.[137] The *Lodi Sentinel* and the *Pacific Rural Press* criticized public enterprises as inefficient, corrupt, or wasteful, but celebrated the announcement that a power utility and two irrigation districts planned to cooperate on a Stanislaus River reservoir.[138] The defeat of the Water and Power Act certainly reflected aggressive campaigning by the state's biggest utilities, but also the deep suspicions of public enterprise that persisted despite the Progressives' political accomplishments and Californians' habitual demonizing of the private utilities.

Until state and federal water development temporarily drowned water disputes in abundance, the advent of EBMUD and other urban plans to divert water from agricultural regions heightened competition for limited resources and, in the process, inspired a new form of antiurbanism. Even the most biting of Lodi's criticisms lacked the bitterness expressed by the communities endangered by the Wachusett Reservoir, however. In part, this difference reflected the diminished expectations of public enterprise. By the 1920s, waterworks no longer offered a mechanism for social engineering, so EBMUD and projects like it carried far less moral baggage than had their nineteenth-century antecedents. Nevertheless, as protection for irrigation in California's water policies demonstrates, rural life still held special significance. Agriculture's moral claims bolstered political and economic arguments and continued to insinuate that urban growth was dangerous and slightly illegitimate. This, of course, did not prevent the transfer of water to urban and industrial uses. It merely encouraged Californians to create new agencies with greater territorial reach and that relied on ever more aggressive engineering to supply water to all claimants.

Conclusion

In 1915, Upton Sinclair published *The Jungle,* a story of the crushing poverty and horrifying working conditions confronting immigrants in Chicago. The metaphor in his title would have been familiar to his readers because urban reformers had been alluding to America's cities as dangerous, disorderly, and fearsome wildernesses for decades. Although proponents of regionalism in Boston and the East Bay never employed this imagery as explicitly as Sinclair did, they shared not only this image of the city but also the conviction that government initiatives could tame the urban jungle. Moral-environmentalists like those active in Boston saw a direct connection between services on the one hand, and urban social and political order on the other. In their minds, a cleaner, orderly city would breed a healthier, more law-abiding and moral population. In the East Bay, Progressives identified private utilities that exploited residents, corrupted elected officials, and endangered urban democracy as the primary threat to the city. Of course, regional water and sewerage could not meet these inflated expectations, but they did transform urban politics and environments. In other words, regionalism built bigger and better jungles.

Before regionalism, water and sewer services in both Boston and Oakland had disappointed residents for decades. Boston could not control the pollution from neighboring communities that contaminated its reservoirs and its waterfront. East Bay municipal officials fretted over poor water service, high water prices, and the ability of water companies to resist regulatory initiatives. Meanwhile, increased competition stymied efforts to satisfy rising water consumption throughout both regions. In Boston, fear that poor sanitation caused epidemics justified aggressive public action to improve drainage and tap clean water supplies. After the germ theory eclipsed moral-environmentalism, as it did by the time East Bay residents contemplated public water development, advocates of public services had to proceed without the benefit of this clear mandate for government ownership. Instead, they argued that only publicly funded improvements could ensure future prosperity. But East Bay residents did not agree on how much their officials should interfere with private enterprise in the pursuit of this goal. So public authority long remained circumscribed in the East Bay.

In both communities, the limitations of municipal government greatly added to the appeal of regionalism even though territorial constraints on municipal authority caused more serious service problems than did corruption or inept administration. By the late nineteenth century, most urbanites recognized that the ways in which municipal boundaries crisscrossed watersheds interfered with sewerage and water supply improvements. Significantly, this realization led city residents, their elected representatives, and urban reformers all to the same conclusion: multiple-city water planning could eliminate both inefficient competition and mutually exclusive water uses. All three groups recognized that interlocal or metropolitan projects could decrease the number of water or sewer systems needed in any urban area while increasing the funds available for public works construction. Moreover, regionalism gave cities access to natural resources that they could not develop independently and assisted them in their efforts to control utility companies. So, although Boston and East Bay residents defined their environmental crises in different terms and endowed city officials with different levels of authority over water and sewer services, regionalism triumphed in both communities because it promised to solve the major deficiencies of municipal enterprise.

Reformers, whether Yankee Republicans in Boston or Progressives in the East Bay, seized upon the political opportunities presented by widespread public dissatisfaction with municipal services and private utilities. They used environmental crises to wrest public works from municipal hands and take control of the new regional agencies. Reformers justified the fact that special districts operated without voter review by citing the need for efficiency and scientific management. But the results were not nearly as apolitical as the rhetoric implied. Rather, special districts were part of an aggressive campaign that included nonpartisan elections, short ballots, civil service statutes, and the weakening of city councils, all intended to transform urban politics. Of these, regionalism was a uniquely potent engine for political reform because it redistributed political power without significantly changing municipal institutions. By removing public works development and the distribution of jobs and services from the hands of elected officials, reformers undermined the patron-client relationships central to the political machine that had given working-class and immigrant voters considerable influence in municipal politics. In other words, regionalism used service improvements to marginalize opponents who had used similar projects as the basis for their power, and regionalism did so largely with the blessing of the very voters and officials disenfranchised by the new institutions.

Outside the urban core, suburban communities, which had less money to spend on public works, needed regional solutions even more than did central cities. Thus, suburbs, too, welcomed special districts because they provided services and preserved home rule. Here too, however, voters and elected officials yielded control over utilities. Ironically, then, although reformers may have intended regional districts to undermine the power of their political opponents, the urban bosses, they also curbed the power of their political allies who dominated suburban government.

The irony of the political transformations brought about by regionalism are even more profound outside the metropolitan area. Grand water diversions like those built by Boston and the East Bay sacrificed rural communities for urban growth. The small towns located near crucial water supplies were keenly aware of this fact. The communities that lost water, land, and other resources resented the limits placed on their own aspirations. They viewed as illegitimate the extension of urban administration so far beyond city boundaries and attacked the cities as corrupt and predacious.

Rural reactions to regional construction revealed a suspicion of cities already deeply rooted in American thinking, which accepted rural and small-town life as the model of virtuous, orderly society. Indeed, the critique of cities as filthy, corrupt, and dominated by foreigners underlay the very urban reform efforts out of which regionalism sprang. From the beginning, regional public works were promoted as a means to cleanse the city and to eliminate scarcity, filth, and squalor. The same aspirations for urban improvements manifested themselves in late-nineteenth- and early-twentieth-century urban planning and public parks movements. Even when moral-environmentalism itself faded, public improvements still promised to create shining, prosperous, orderly cities. Even though regionalism did not accomplish all these goals, metropolitan systems did clean the cities and provide ample water supplies. By so doing, they removed environmental barriers to urban expansion. The physical and administrative expansion of metropolitan areas, then, took place on the backs of the very rural communities that had for years embodied the American ideals of civil society.

Regionalism's Longer Pipe

From the perspective of the 1890s or the 1920s, regional public works appeared to be an unalloyed success. Rural reservoirs relieved water shortages, decreased contamination and disease, and facilitated fire protection and industrial growth. Regional planning eliminated both haphazard municipal public works construction and the interlocal resource competition that had interfered with so many municipal water and sewer projects. Coordinated sewerage removed wastes from urban shorelines and prevented the sewage pollution that had long plagued low-lying urban areas. Contemporary observers also believed that regional administration eliminated municipal graft and influence-peddling. But despite the fanfare that greeted the completion of pipes, pump stations, dams, and aqueducts, regionalism did not permanently solve urban public service problems. Eventually, the environmental crises returned. By facilitating continued expansion, new water supplies allowed cities to outgrow even the most ambitious regional networks. Meanwhile, sewage silently accumulated in deeper waters. Rather than fundamentally changing urban relationships to natural resources, in effect regional networks merely built longer pipes.

When regional special districts were first approved, their political innovations were not met with similar changes in engineering or resource development policy. This is what is meant by the description of regional systems as merely longer pipes. The physical structures built by individual cities or private water companies and by metropolitan agencies had much in common, in part because cities outgrew local resources before technology or science provided new techniques for satisfying urban service needs. But regional networks also retained the characteristics of their municipal and private antecedents because regional officials saved enormous amounts of time and money by simply grafting bigger sewer mains and water supplies onto existing service networks. The adaptation of existing infrastructure physically locked regional officials into established water supply and sewerage practices.[1] As a result, building reservoirs and sewers eliminated neither water supply nor waste disposal problems, but merely moved or delayed them.

The history of waterworks in Boston clearly demonstrates the ways in which successive service improvements prompted the construction of ever-longer aqueducts. Every time Boston ran out of water, municipal and then regional officials sought new reservoirs. From local wells, these networks extended first to Lake Cochituate, then to the Sudbury River, and then to Wachusett Reservoir. In the 1920s, Massachusetts regional officials extended the water system again, returning to western Massachusetts for more water only a few years after the completion of the Wachusett Reservoir and long before metropolitan Boston faced imminent water shortages. They diverted the Ware River into Wachusett Reservoir and dammed the Swift River to form Quabbin Reservoir. Approved in 1927 and completed in 1939, Quabbin drowned four towns and four cemeteries, and dislocated twenty-five hundred people.[2] Residents of Enfield, Dana, Greenwich, and Prescott repeated almost verbatim the objections that their Nashua Valley counterparts had raised in the 1890s, but with no greater success.

Boston's victory over the Swift Valley towns created lasting bitterness between eastern and western sections of the state. Since the 1980s, the Boston area has again begun to experience water shortages. In a notable break from the past, however, neither its regional authority nor the public have demanded additional supplies. Instead, regional officials have embarked on an aggressive campaign to plug leaks in aqueducts and water lines and urge consumers to conserve water. They expect system repairs alone will increase available water by as much as twenty-five percent. The vivid memory of the destruction that accompanied the building of Quabbin Reservoir has rendered politically impossible, at least for the foreseeable future, any diversion of the Deerfield, Westfield, or Connecticut Rivers to the metropolitan area. In Massachusetts, then, sectional rivalry replaced interlocal resource competition, and may finally have placed real limits on further urban resource appropriation.

The East Bay Municipal Utility District has had greater success in extending its water system to keep pace with growth, not because the district has built new reservoirs or grown more slowly than Boston, but because California's aggressive water development has created additional water options for urban communities. State and federal agencies have built massive irrigation networks in California.[3] In the name of prosperity and progress, California's two largest water systems, the Central Valley Project and State Water Plan, turned nearly every river in the state to agricultural, industrial, or domestic uses. Beginning in 1970, EBMUD and other cities began purchasing water from the federally operated Central Valley Project.[4] Although this practice has come under fire, these water purchases have given the East Bay access to an apparently limitless water supply and has continued the long history of urban water diversions from agriculture.

More than half of the water used in California flows through the Central Valley Project and State Water Plan facilities; agriculture consumes eighty-five percent of this. Over eighteen hundred small, private projects withdraw an additional third of the total flow from the Sacramento and San Joaquin rivers. Diversions from this watershed are so extensive that, during dry years, pumps draw water upstream through much of the Delta to send it south to irrigate the San Joaquin Valley. State and federal officials plan to increase diversions from the Sacramento–San Joaquin watershed by a third over the next decades.[5] Such development has devastated California fisheries and wildlife habitat and removed decision making about river use from local hands.

Despite continued thirst for urban and agricultural water, California's aggressive water development has come under increasing fire because it has so disrupted the natural environment and infringed on local control over natural resources. Because of its devastating implications for San Francisco Bay, voters in the 1970s defeated a proposed peripheral canal that would have diverted millions of gallons of water from the northern Sacramento Delta to the southern San Joaquin Valley.[6] Los Angeles' attempts to take water from Mono Lake's tributaries sparked successful opposition in the 1980s.[7] In 1991, Congressman George Miller sponsored a federal bill to require the Central Valley Project administrators to release 800,000 acre feet of water into natural riverbeds to assist in fisheries and habitat restoration.[8] In an earlier era, California law would have defined the unused water as an unacceptable waste of valuable resources because state water policies allocated water to only the narrowest range of human endeavors. But, as these examples demonstrate, increasing numbers of Californians have begun to question the development orientation imbedded in these policies.

Regional sewage presents an even more graphic demonstration than water supply of regionalism's longer pipes. The identification of germs as the cause of disease permitted cities to focus their sewerage efforts narrowly on blocking the transmission of germs. Furthermore, their preoccupation with human disease permitted public officials to ignore industrial effluent or even to advocate unrestricted disposal of manufacturing wastes because harsh chemicals were thought to kill disease-bearing organisms and purify domestic sewage.[9] So, where nineteenth-century sanitarians had addressed a broad range of urban conditions, their twentieth-century successors now considered adequate any project that minimized human contact with infectious materials either by disinfecting sewage or by transporting wastes to deep waters. Both Boston and the East Bay chose the latter option. The exclusive focus on human disease delayed implementation of treatment and disposal methods that might have prevented the long-term environmental damage caused by industrial waste disposal in San Francisco Bay and Boston Harbor.

Criticism of the narrow goals of twentieth-century sewage disposal first emerged after World War II, when accumulated pollution along metropolitan shores began to interfere tangibly with fishing, recreation, and other waterfront activities. In the 1950s, Boston's consolidated regional water, sewer, and park agency, the Metropolitan District Commission, built swimming pools to replace polluted beaches, and erected a plant on Nut Island to treat wastes from the southern metropolitan drainage system. In 1968, a similar facility on Deer Island began filtering effluent from northern pipes. Unfortunately, because of malfunctioning pumps and other technical difficulties, the plants often operated at less than half capacity. Excess sewage, diverted to the Moon Island outfalls built in the 1880s for the Main Drain, poured into the inner harbor. By the 1980s, the District's commitment to minimal treatment, increased volumes of waste water, and three centuries of harbor disposal had not only overwhelmed the aging treatment plants, but also the harbor's capacity to absorb organic and inorganic wastes.

By 1983, over five billion gallons of raw and minimally treated sewage were spilling into Boston Harbor annually. Wastes from nearly two million people left a trail of fecal matter, "tampon applicators, condoms, grease and oil" across harbor beaches.[10] Together with sludge from the treatment plants themselves, this sewage regularly forced officials to close Boston's nineteen miles of public beaches and two thousand acres of shellfish beds.[11] Untreated sewage also escaped into the harbor from over a hundred combined sewer overflows along the waterfront. Originally intended to keep sewage from flooding city streets during the heaviest rains of the year, the sewer overflows dated from the 1880s. A hundred years later, household and commercial wastes filled or overflowed Boston's sewer mains in some sections of the city even during droughts, so untreated effluent regularly spewed directly into the harbor.[12] These conditions were all too similar to the ones that had inspired the construction of the Main Drain and the organization of the Metropolitan Drainage Commission in the nineteenth century.

Early in 1983, a morning walk on a beach covered in sewage, grease, and debris from the Moon Island outfall inspired William Golden, then

Quincy Town Solicitor, to sue the Metropolitan District Commission for polluting Boston harbor.[13] At one point, Quincy even requested an injunction that would have outlawed any new sewer connections in the metropolitan Boston area until the district improved its sewage disposal practices. Eventually, State Supreme Court Justice Paul Garrity ordered the district to renovate its treatment plants, reduce the amount of sewage flowing into the system, eliminate the combined sewer overflows, and implement secondary treatment.[14] Garrity's ruling essentially forced the district to implement the water quality standards laid out in the federal Clean Water Act of 1972. Because the Metropolitan District Commission failed to comply, in 1984 the legislature created a new agency, the Massachusetts Water Resources Authority, to carry out the Boston Harbor Cleanup.[15] This agency has since completed a new secondary treatment plant on Deer Island, a nine-mile outfall into Massachusetts Bay, and built a storm-water storage facility to prevent the discharge of raw waste from combined sewer overflows. Federal rules now bar metropolitan Boston from dumping sewage sludge in the ocean. As a result, the Massachusetts Water Resources Authority strictly regulates hazardous and toxic chemicals in regional sewerage so it can manufacture usable fertilizer from the sludge.[16] Although the Boston Harbor Cleanup does, in some ways, represent a longer pipe by extending regional sewerage outfalls into yet deeper water, the provisions to limit total sewage volume and reduce chemical pollutants promise to do far more to protect the marine environment than did earlier sewerage policies.

In recent decades, East Bay officials have also started to abandon established sewage disposal strategies. In this case, continued water shortages rather than complaints regarding sewage pollution have provided the impetus for innovation. In 1972, the Environmental Defense Fund challenged EBMUD's claim on irrigation water in the Central Valley Project. The group argued that increased urban withdrawals from the irrigation system would divert 134 million gallons of water that the natural systems of the Delta could ill afford to lose. In 1990, the Alameda County Superior Court ruled that EBMUD could continue to tap the Central Valley Project, but only when water in the Delta exceeded minimum levels. Finding their use of Central Valley Project water so curtailed, East Bay regional officials have pinned their hopes on wastewater recycling. In 1991, they began supplying specially treated sewage water to three golf courses and an oil refinery in the East Bay. Several Bay Area advocacy groups are backing proposals to collect sewage from the entire Bay region for irrigation use in the San Joaquin Valley.[17] These plans represent an even greater departure from the old dilute and dispose strategy than does the Boston Harbor Cleanup. Where regionalism merely extended service lines, recycling, conservation, and limits on chemical pollutants may represent real changes in urban resource policies.

Since the inception of urban services as municipal and private initiatives, city dwellers have built longer pipes to satisfy their need for services. Regionalism continued this trend, giving urban residents unprecedented access to distant resources. Source communities, particularly those considered economically marginal, had little ability to block the plans of larger, wealthier metropolises. But, just as nineteenth-century urban annexation created a home-rule backlash, metropolitan districts' repeated victories have deepened resistance to further urban resource exploitation. Since the 1960s, an unlikely coalition between source communities and a substantial minority of urban voters has challenged both the ambitions of regional agencies and urban expansionism. In Massachusetts, this coalition has grown powerful enough to prevent new reservoir construction for Boston and to force metropolitan authorities to undertake the Boston Harbor Cleanup. In California, the combination of source community and environmental activism has not only curbed new water development but also returned water to rivers for the use of fish and other wildlife.

Members of these coalitions have come to challenge regionalism's ever-larger water and sewer networks because of their frustration that metropolitan public works have neither satisfactorily addressed rural needs nor accommodated the changing priorities of urban voters. This isolation from voter influence was deliberate and initially greeted with much acclaim. In the eyes of the advocates of regionalism, bureaucracy and expertise ensured that efficiency and scientific principles, not political expediency or favoritism, would determine water supply and sewerage designs. This same political isolation, however, allowed engineers to pursue very circumscribed policy objectives and ignore changes in public priorities for regional systems. In other words, their political isolation enabled regional agencies to continue their established patterns of building longer pipes in response to all water and sewerage demands, irrespective of changing circumstances or public opinion.

Impelled by evidence of environmental decline, new critics have questioned Boston and East Bay regional agencies' assumptions about water resources. These critics have moved to claim water resources for activities and individuals undervalued in the pro-growth climate of the Gilded

Age and Progressive Era. Environmentalists seek to increase the amount of water available to preserve biodiversity, wild rivers, unique aquatic ecosystems, and fisheries. Other activists want to redress the inequitable distribution of resources that denied rural water supplies to people based on class, occupation, or ethnicity.[18] The new critics of regionalism are engaged in the same deeply entrenched critique of urban development that spawned but can no longer be contained within regionalism.

Environmentalists in particular have called attention to the long-term implications of regional water and sewer policies. Descriptions of environmental damage have emerged in discussions about metropolitan public works only as urban residents have seen the unanticipated consequences of regionalism. In fact, since the 1960s, environmental collapse has overshadowed both drought and public health in discussions of regional water management. The language of environmental crisis has grown to encompass an increasingly varied set of demands for resources. Meanwhile, in many discussions of urban services fear of environmental devastation has replaced fear of economic, political, or social crisis as a primary rationale for changes in public policy. This shift has focused new attention on fish and aquatic ecosystems as valuable resources, and has given recreation and aesthetic qualities new status in the water distribution equation. This trend reflects the greater influence of attitudes that either had little influence or did not exist at the time Boston and the East Bay adopted regionalism.

Despite their growing influence, the new critics of urban resource policy remain largely excluded from metropolitan agencies. As a result, they have sought to influence regional officials from the outside. Proponents of the Boston Harbor Cleanup, for example, used the judicial system to force sewer officials to hear the opinions of those who favored the cleanup, a majority that it had largely ignored. The Boston Harbor Cleanup, like the Clean Water Act and other federal environmental regulations, was instituted in response to intensive grassroots lobbying by environmental groups that believed that local or state agencies had become too identified with commercial and development interests. The appeal to federal authorities mirrored the efforts of an earlier generation of reformers to overcome the limitations of municipal governments by creating regional special districts. The litigation and lobbying that resulted in federal environmental regulation represented an important means by which excluded constituencies could influence water resource policy.

Had regional agencies altered their water and sewer policies in response to evolving public demands, they might have staved off some criticism and litigation. However, the institutional structures erected to protect regional public works from graft and partisan wrangling also isolated regional agencies from changing conditions and values within the metropolitan communities. Moreover, once built, water and sewer systems committed Boston and the East Bay to continue to implement outmoded conceptions of water supply and sewerage long after the public had embraced newer visions of water resources. The inflexibility of water and sewer systems to accommodate new policy priorities and new conceptions of water resources magnified the effects of excluding so many voices from water policy discussions.

A Bigger Jungle

The regional networks constructed for Boston and the East Bay have been unquestionably successful by several measures. They gave to growing cities the water supplies and waste disposal mechanisms they needed to prosper. By expanding public authority over natural resources they increased public access to vital services. They also eliminated the interlocal conflict that had hindered water supply and sewerage development for decades. On the other hand, these systems intensified and concentrated resource use in some potentially dangerous ways. Narrow policy goals and extensive water use devastated rivers, marshes, and bays. Overconfidence in the ability of technology to meet resource needs by manipulating natural systems ultimately impaired the ability of those natural systems to absorb the byproducts of modern industrial life. Faith in engineering expertise and isolation from voters at once blinded regional agencies to the ecological implications of urban development and discouraged critical analysis of urban resource policy.

At the inception of regionalism, localized environmental problems caused by inadequate water supply and sewage disposal overwhelmed Boston and the East Bay cities. Since the 1960s, the problems have reemerged, but on a vastly greater scale. The resulting reconception of environmental crisis as extending beyond the metropolis has spawned a new generation of superregional agencies to administer environmental protection in territories that extend over thousands of square miles. Water supply battles now involve multiple states, as the Colorado River

Compact demonstrates. Controlling sewage pollution likewise requires coordination in territories far larger than an isolated metropolis. The San Francisco Bay Conservation and Development Commission regulates landfill, development, and water quality for the entire estuary, for example. In the East, the Massachusetts Bays Program coordinates local, state, and federal initiatives, while Chesapeake Bay and Great Lakes management require the participation of several states. Unlike the original metropolitan special districts, however, these superregional agencies rarely have adequate enforcement powers or independent sources of funding. Thus they cannot act as aggressively as the metropolitan agencies once did.

The expansion from regional to superregional agencies has reinforced many of the original political changes associated with the metropolitan special district. Institutions that control whole watersheds or estuary systems centralize resource policies even more than did EBMUD or the Metropolitan Water Board. Although few people discuss the implications of these agencies for home rule, their size further removes water resource policies from public influence. The new agencies do tend to consult experts from a wider range of fields—including marine and fisheries biologists—in addition to hydraulic engineers. But their efforts to incorporate wildlife management and environmental quality into their policies have made superregional agencies even more dependent than their predecessors on technical expertise and centralized government power. Ultimately, however, the success of these superregional agencies will depend on their ability to adjust to public demands for resources and to keep their conception of environmental problems inclusive and flexible. Centralization may make sense from a purely ecological sense because it permits administration of whole watersheds as well as effective mediation of conflicting resource demands. On the other hand, superregional agencies may ultimately define resources and policy debates as narrowly as metropolitan special districts have over the past century. Coordination of development across larger territories appears promising and may indeed be the only option for interstate rivers and bays such as the Colorado River and the Gulf of Maine. But superregions may also permit communities to overlook very real limits on continued growth and thus may spread the hazards of urban development yet further into the surrounding countryside.

Notes

Introduction

1. During the 1988 presidential campaign, then Vice President Bush aired a campaign ad that implied that Dukakis, as Massachusetts governor, bore responsibility for Boston Harbor pollution. The ad called the harbor the most polluted in the country and glossed over the fact that the Reagan administration had decreased federal aid that might have helped clean the harbor sooner. The ad also contributed to Bush's efforts to paint himself as an environmental candidate.

2. Metropolitan special districts are also known as authorities, boards, and commissions. Although different terms occasionally denote the ways in which agency officials are appointed or elected, they have been used interchangeably in most communities and by most historians. Semiautonomous agencies are found at all levels of government. Where necessary, I have distinguished between *municipal* agencies operating in a single city and *regional* or *metropolitan* agencies with jurisdiction in several cities or counties.

3. Robert Higgs, in *Crisis and the Leviathan: Critical Episodes in the Growth of American Government* (New York: Oxford University Press, 1987), and Stephen Skowronek, in *Building a New American State: The Expansion of National Administrative Capacities, 1877–1920* (New York: Cambridge University Press, 1982), argue that temporary, acute crises cause permanent government expansion.

4. The spread of innovations in municipal administration, as described by Thomas Scott, "The Diffusion of Urban Government Forms as a Case of Social Learning," *Journal of Politics* 30 (1968): 1091–1108, offers one model for the way cities adopted proven technologies.

5. On the importance of new technologies in advancing urban sanitation, see, for example, Joel Tarr and Gabriel Dupuy, eds., *Technology and the Rise of the Networked City in Europe and America* (Philadelphia: Temple University Press, 1988); and Joel Tarr, "The Separate versus Combined Sewer Problem: A Case Study in Urban Technology Design Choice," *Journal of Urban History* 5 (1979): 308–40.

6. Nancy Burns, *Formation of American Local Governments: Private Values in Public Institutions* (New York: Oxford University Press, 1994), pp. 14, 48; Annmarie Hauk Walsh, *The Public's Business: The Politics and Practices of Government Corporations* (Cambridge: MIT Press, 1978), pp. 19–23.

7. On the history of special districts in the United States, see Burns, *Formation of American Local Governments;* and Winston W. Crouch and Dean E. McHenry, *California Government: Politics and Administration* (Berkeley: University of California Press, 1954).

8. See Thomas H. O'Connor, *The Boston Irish: A Political History* (Boston: Northeastern University Press, 1995), pp. 95–99; Ronald P. Formisano and Constance K.

Burns, eds., *Boston, 1700–1980: The Evolution of Urban Politics* (Westport, Conn.: Greenwood Press, 1984); and Adelaide M. Cromwell, *The Other Brahmins: Boston's Black Upper Class, 1750–1950* (Fayetteville: University of Arkansas Press, 1994).

9. For the ethnic, economic, and political origins of antimonopoly sentiment in California, see Spencer C. Olin, Jr., *California's Prodigal Sons: Hiram Johnson and the Progressives, 1911–1917* (Berkeley: University of California Press, 1968); William Deverell, "Building an Octopus: Railroad and Society in Late Nineteenth Century." Ph.D. diss., Princeton University, 1989; and Mansel G. Blackford, *Politics of Business in California, 1890–1920* (Columbus, Ohio: Ohio State University Press, 1977).

1. Municipal and Private Services in the Nineteenth Century

1. John Koren, *Boston, 1822 to 1922: The Story of Its Government and Principal Activities During One Hundred Years* (Boston: 1923), pp. 6, 8, 161.

2. For a discussion of changing attitudes toward individual responsibility for disease, see the classic Charles Rosenberg, *The Cholera Years: The United States in 1832, 1849 and 1866* (Chicago: University of Chicago Press, 1962). See also Christopher Hamlin, *A Science of Impurity: Water Analysis in Nineteenth Century Britain* (Berkeley: University of California Press, 1990) and Bill Luckin, *Pollution and Control: A Social History of the Thames in the Nineteenth Century* (Bristol, England: Hilger, 1986). Judith Walzer Leavitt explains the reactions of the poor and immigrants to quarantines and other public health measures in *The Healthiest City: Milwaukee and the Politics of Health Reform* (Princeton, N.J.: Princeton University Press, 1982). See also Martin Pernick, "Politics, Parties and Pestilence," in Judith Walzer Leavitt and Ronald L. Numbers, eds., *Sickness and Health in America: Readings in the History of Medicine and Public Health* (Madison: University of Wisconsin Press, 1985), pp. 356–71.

3. Edwin Chadwick, *Report on the Sanitary Condition of the Labouring Population of Great Britain* (1842).

4. Rosenberg, *The Cholera Years,* pp. 4–7.

5. Donald Reid has explored this association between unsanitary conditions and social upheaval, describing how Parisians associated their sewers with revolution and with social outcasts. Donald Reid, *Paris Sewers and Sewermen: Realities and Representations* (Cambridge, Mass.: Harvard University Press, 1991), pp. 2–4, 14–15, 18–35.

6. "The Sewerage Necessity," *Boston Morning Journal,* 1 August 1877.

7. "Sanitary Reform" in *Edinburgh Review* 91 (1849): 213, 218, reprinted in *Public Health in the Victorian Age: Debates on the Issue from Nineteenth Century Critical Journals* (Westmead, England: Gregg International, 1973).

8. Boston, *Ordinances Prescribing Rules and Regulations Relative to Nuisances, Sources of Filth, and the Causes of Sickness within the City of Boston* (Boston: 1833), p. 1.

9. Joel A. Tarr has written a number of pivotal works on the history of technology and urban sanitation. These include "The Development and Impact of Urban Wastewater Technology: Changing Concepts of Water Quality Control, 1850–1930," in Martin V. Melosi, ed., *Pollution and Reform in American Cities, 1870–1930* (Austin: University of Texas Press, 1980), pp. 59–82, and "Sewerage and the Development of the Networked City in the United States, 1850–1930," in Joel A. Tarr and Gabriel

Dupuy, eds., *Technology and the Rise of the Networked City in Europe and America,* (Philadelphia: Temple University Press, 1988), pp. 159–85, among others. Stuart Galishoff has studied sanitation in both Newark and Atlanta. See, in particular, "Drainage, Disease, Comfort and Class: A History of Newark's Sewers," *Societas: A Review of Social History* 6 (1976): 121–39, and *Newark: The Nation's Unhealthiest City, 1832–1895* (New Brunswick, N.J.: Rutgers University Press, 1988). Andre Guillerme's investigations of French sanitation, "The Genesis of Water Supply, Distribution and Sewerage Systems in France, 1800–1850" and Georges Knaebel's studies of Bielefeld, Germany, "Historic Origins and Development of a Sewerage System in a German City," both in *Technology and the Rise of the Networked City,* pp. 91–115, 186–206, demonstrate that similar problems afflicted cities in Europe.

10. Boston, *Digest of Laws and Ordinances Relating to the Public Health* (Boston: 1869), pp. 38–39.

11. See, for example, Massachusetts Medical Commission, *The Sanitary Condition of Boston: The Report to the Boston Board of Health* (Boston: 1875), pp. 33, 62–68.

12. U.S. Department of Interior, Census Office, "Report on the Social Statistics of Cities," *Tenth Census of the United States,* vol. 18 (Washington, D.C.: 1886), p. 130.

13. As late as 1885, many members of the Boston city government planned to recoup three-quarters of sewer construction costs through special property tax assessments levied on those requesting and benefiting from new sewer construction. Hugh O'Brien to Boston Aldermen, 20 July 1885, in *Reports of Proceedings of the City Council of Boston* (Boston: 1885), p. 467 [hereafter cited as *City Council Proceedings*].

14. Dr. C. E. Buckingham to Josiah Curtis, 7 December 1848 in Josiah Curtis, *Brief Remarks on the Hygiene of Massachusetts, but More Particularly the Cities of Boston and Lowell, Being a Report to the American Medical Association* (Philadelphia: 1849), p. 16.

15. Robin Einhorn observed a similar distribution of public works by class in *Property Rules: Political Economy in Chicago, 1833–1872* (Chicago: University of Chicago Press, 1991).

16. Dr. C. E. Buckingham to Josiah Curtis, 7 December 1848, in Curtis, *Brief Remarks,* p. 15.

17. William Ripley Nichols and George Derby, *Sewerage; Sewage; the Pollution of Streams; the Water-Supply of Towns: A Report to the State Board of Health of Massachusetts* (Boston: 1873), pp. 32–33; Boston, *Digest of Statutes and Ordinances Relating to the Public Health* (Boston: 1873), pp. 33–34.

18. Donald W. Howe, *Quabbin, the Lost Valley* (Ware, Mass.: Quabbin Book House, 1951), pp. 6–7; Nathaniel J. Bradlee, *History of the Introduction of Pure Water into the City of Boston* (Boston: 1868), p. 1. Although the 1676 fire does not seem particularly devastating to modern eyes, at the time Boston contained only 1,719 homes.

19. Josiah Quincy, "Address to the Board of Aldermen and Members of the Common Council . . . January 2, 1826," in *The Inaugural Addresses of the Mayors of Boston,* vol. 1 (Boston: 1894), pp. 52–53; John C. Warren to Josiah Quincy, 25 November 1825, in John C. Warren Papers, vol. 3, Massachusetts Historical Society, Boston, Massachusetts; Howe, *Quabbin,* pp. 7–8.

20. George Odiorne served on the Boston Board of Aldermen in 1823 and as Suffolk County's senator in the General Court in 1824. The fact that his colleagues on the city council consistently refused his water proposal may be seen as further evi-

dence of their commitment to public water development. See Bradlee, *History of the Introduction of Pure Water*, p. 15; Robert H. Eddy, *Report on the Introduction of Soft Water into the City of Boston* (Boston: 1836), pp. 34–35.

21. Laommi Baldwin, *Report on the Subject of Introducing Pure Water into the City of Boston* (Boston: 1834), pp. 37–38

22. Daniel Treadwell, "Report Made to the Mayor and Aldermen of the City of Boston, on the Subject of Supplying the Inhabitants of that City with Water" (Boston: 1825), pp. 3–5, 9, in *Municipalities: Water, Gas, Street Railways*, pamphlet collection, Widener Library, Cambridge, Massachusetts; Baldwin, *Report on the Subject of Introducing Pure Water*, pp. 39–40. At the time, Cochituate was known as "Long Pond." The lake was renamed Cochituate after Boston began using it as a reservoir. I have used this later name throughout.

23. Baldwin, *Report on the Subject of Introducing Pure Water*, pp. 39–45.

24. Bradlee, *History of the Introduction of Pure Water*, pp. 2–3; Theodore Lyman, Jr., *Communication to the City Council, on the Subject of the Introduction of Water into the City* (Boston: 1834), p. 24. In 1843, the Council approved a proposal for a new fire reservoir in South Boston. This reservoir was designed to serve only neighborhood structures. "Records of the City of Boston Mayor and Aldermen, 1840," p. 204, manuscript record books, uncatalogued Boston City Records collection, Boston Public Library, Boston, Massachusetts. (Hereafter cited as "Aldermen Records.")

25. Massachusetts General Court, Joint Special Committee to Which Was Referred the Petition of Boston . . . to Introduce into Boston, Pure, Soft Water from Long Pond, "Report," 13 March 1845, 8, no. 92, in *Documents Printed by Order of the Senate of the Commonwealth of Massachusetts* (Boston: 1845).

26. "Aldermen Records, 1835," p. 176.

27. Theodore Lyman, Jr., "Address Made to the City Council of Boston, 5 January 1835," in *The Inaugural Addresses of the Mayors of Boston*, vol. 1 (Boston: 1894), pp. 180, 202–3; "Inaugural Address of Samuel T. Armstrong, January 4, 1836," in *The Inaugural Addresses of the Mayors of Boston*, vol. 1 (Boston: 1894), p. 208.

28. Massachusetts General Court, "Report," p. 6. Altogether, 2,107 voted for municipal water; only 136 meeting attendees opposed the project.

29. Samuel A. Eliot, "Inaugural Address," 2 January 1837; "Address of the Mayor to the City Council of Boston," 1 January 1838; "Address of the Mayor to the City Council of Boston," 7 January 1839," in *The Inaugural Addresses of the Mayors of Boston*, vol. 1 (Boston: 1894), pp. 211, 222–23, 232.

30. Samuel A. Eliot, "Address of the Mayor to the City Council of Boston," 1 January 1838, and "Address of the Mayor to the City Council of Boston," 7 January 1839," in *The Inaugural Addresses of the Mayors of Boston* vol. 1 (Boston: 1894), pp. 223, 232; "Aldermen Records, 1838," p. 120. Most voters (2,541 of 4,162) answered affirmatively the question "Is it expedient for the City to procure a supply of soft water, at its own expense?" Somewhat fewer (2,507) felt it expedient to begin work in 1838 if the legislature were to grant proper authority.

31. Jonathan Chapman, "Address of the Mayor to the City Council of Boston," 6 January 1840, in *The Inaugural Addresses of the Mayors of Boston*, vol. 1 (Boston: 1894), pp. 247, 250.

32. "Return of Votes from the Several Wards on Water, December 9, 1844," in *Return*

of Votes for the City of Boston, November 9 1840–March 1853, manuscript vote records, Boston Public Library, Boston, Massachusetts; Massachusetts General Court, "Report," 13 March 1845, pp. 8–9; Thomas A. Davis, "The Mayor's Address to the City Council of Boston," 27 February 1845, in *The Inaugural Addresses of the Mayors of Boston*, vol. 1 (Boston: 1894), pp. 316–17.

33. John H. Wilkins, "Remarks on Supplying the City of Boston with Water" (Boston: 1845), p. 4, in *Municipalities: Water, Gas, Street Railways*, pamphlet collection, Widener Library. Emphasis his.

34. Walter Channing, "A Plea for Pure Water" (Boston: 1844), in *Boston Water Supply 1844–1845* pamphlet collection, Widener Library; John C. Warren to Josiah Quincy, 25 November 1825, in John C. Warren Papers.

35. "Aldermen Records, 1844," pp. 169–71.

36. "Aldermen Records, 1846," pp. 373.

37. "Remonstrance of A. L. Brooks and 190 Others . . ." and "Remonstrance of the Inhabitants of Framingham . . ." February 1845, in Legislative Packet for Chapter 220, General Court Acts of 1845, Massachusetts Archives, Boston.

38. "Remonstrance of the Proprietors of the Middlesex Canal . . ." 28 January 1845, in Legislative Packet for Chapter 220, General Court Acts of 1845, Massachusetts Archives, Boston.

39. "Aldermen Records, 1840," pp. 12, 13.

40. "Letter from Lemuel Shattuck in Answer to Interrogatories of J. Preston in Relation to the Introduction of Water into the City of Boston" (Boston: 1845), pp. 27–30, in *Municipalities: Water, Gas, Street Railways*, pamphlet collection, Widener Library; "Thoughts about Water" (Boston: 1844), pp. ii, 4, 6; William J. Hubbard, "Mr. Hubbard's Argument before the Joint Special Committee of the Massachusetts Legislature on the Water Question" (Boston: 1845), p. 5, in *Boston Water Supply 1844–1845*, pamphlet collection, Widener Library.

41. "Aldermen Records, 1845," p. 12.

42. "Thoughts about Water" (Boston: 1844), p. 7, in *Boston Water Supply 1844–1845*, pamphlet collection, Widener Library.

43. "Returns of Votes from the Several Wards for or against the Water Act, May 19, 1845," in *Return of Votes for the City of Boston, November 9 1840–March 1853*, manuscript vote records, Boston Public Library, Boston, Massachusetts; Josiah Quincy, Jr., "The Mayor's Address to the City Council of Boston," January 5, 1846, in *The Inaugural Addresses of the Mayors of Boston*, vol. 1 (Boston: 1894), p. 326.

44. "Returns of Votes for the Several Wards on the Water Act, April 13, 1846," in *Return of Votes for the City of Boston, November 9 1840–March 1853*. The final tally showed that only 348 of 4,985 voters opposed the water act.

45. See Einhorn, *Property Rules*, for more on this nineteenth-century transition.

46. Chapter 167, *General Court Acts 1846*.

47. Spencer C. Olin, Jr., *California's Prodigal Sons: Hiram Johnson and the Progressives, 1911–1917* (Berkeley: University of California Press, 1968), pp. 2, 45–50; William Deverell, "Building an Octopus: Railroad and Society in Late Nineteenth Century." Ph.D. diss., Princeton University, 1989, pp. 131–32; Mansel G. Blackford, *Politics of Business in California, 1890–1920* (Columbus, Ohio: Ohio State University Press, 1977), pp. 78–80.

48. Donald Pisani, *From the Family Farm to Agribusiness: The Irrigation Crusade in California and the West, 1850–1931* (Berkeley: University of California Press, 1984), p. 202; Olin, *California's Prodigal Sons,* pp. 41–43; Deverell, "Building an Octopus," pp. 131–32.

49. Ward M. McAfee, "Local Interests and Railroad Regulation in Nineteenth Century California" (Ph.D. diss., Stanford University, 1965), pp., 6, 17, 24; Blackford, *Politics of Business in California,* p. 95.

50. Lois Rather, *Oakland's Image: A History of Oakland, California* (Oakland: Rather Press, 1972), pp. 11–26, 28, 34–37; John Wesley Noble, *Its Name Was M.U.D.* (Oakland: East Bay Municipal Utility District, 1970), p. 2.

51. Ruth Hendricks Willard, *Alameda County, California Crossroads: An Illustrated History* (np: Windsor Publications, 1988), pp. 36–37.

52. "Report of the Board of Engineers on the Grades, Streets and Sewerage of the City of Oakland, 1869" (Oakland: Oakland News, 1870), pp. 11, 12, in *Pamphlets on Oakland,* vol. 1, Bancroft Library, Berkeley, California.

53. "Report of the Board of Engineers on the Grades, Streets and Sewerage," p. 11.

54. W. R. Davis, "Annual Message, 1887," p. 3, in *Pamphlets on Oakland,* vol. 1, Bancroft Library, Berkeley.

55. Anson Barstow, "Annual Message" (Oakland: 1902).

56. Donald Pisani, in *To Reclaim a Divided West: Water, Law, and Public Policy, 1848–1902* (Albuquerque: University of New Mexico Press, 1992), has detailed the origins of western water law and the relationship between water rights and economic and political development in the arid West. According to Pisani, prior appropriation was "the greatest legal innovation in the history of the arid West" because it transformed water from a common good to a marketable commodity. According to the doctrine of riparian rights inherited by the United States from English common law, each landowner along a stream had the same right to water flowing past or through his or her lands. This precluded any single landowner from polluting, diverting, or reducing the volume of a stream so much that it reduced the value of the water to his or her downstream neighbors. Riparian rights did not permit the large-scale water diversions necessary for hydraulic mining and irrigation in the arid West. But the principle of prior appropriation, which separated water rights from land ownership, permitted the expansion of the western economy.

57. Noble, *Its Name Was M.U.D.*, p. 3; flyleaf of George Pardee Scrapbooks, vol. 1, Special Collections, Stanford University Library, Palo Alto, California.

58. Noble, *Its Name Was M.U.D.*, p. 3.

59. S. M. Marks, "From the Beginning: An Outline of the History of the Water Supply for the East Bay District from 1854 to 1910" (1922?), p. 1, in *The History of Water and Water Companies in the East Bay Area,* East Bay Municipal Utility District Records Office, Oakland; S. T. Harding, *Water in California.* (Palo Alto, Calif.: N-P Publications, 1960), p. 119.

60. Noble, *Its Name Was M.U.D.*, p. 4.

61. Noble, *Its Name Was M.U.D.*, p. 7; Marks, "From the Beginning," pp. 1–2.

62. Marks, "From the Beginning," p. 2. The San Leandro reservoir was eventually named after Chabot.

63. "Prospectus of the Oakland and Alameda County Water Company to Furnish the Cities of Oakland and Alameda with Pure Water" (Oakland: 1877), box 4 in *Oak-*

land Miscellany, Bancroft Library, Berkeley. In 1877, the Oakland and Alameda Water Company, inactive since it formed in 1866, proposed to supply the city with water from the San Joaquin River.

64. *Oakland Times* 17 May 1890 and Oakland *Enquirer* 28 January 1890, in George Pardee Scrapbooks, vol. 1, Special Collections, Stanford University Library.

65. Marks, "From the Beginning," pp. 3, 6; Noble, *Its Name Was M.U.D.*, pp. 9, 10.

66. Marks, "From the Beginning," p. 8.

67. Thomas Bowhill, "The Alvarado Artesian Water of the Oakland Water Company Compared with the Surface Waters of Lake Temescal and Lake Chabot of the Contra Costa Water Company" (Oakland: 1895), p. 41, East Bay Municipal Utility District Archives, Oakland.

68. Bowhill, "The Alvarado Artesian Water," p. 42.

69. Noble, *Its Name Was M.U.D.*, p. 10.

70. Noble, *Its Name Was M.U.D.*, p. 10.

71. Noble, *Its Name Was M.U.D.*, p. 10.

72. Noble, *Its Name Was M.U.D.*, p. 11.

73. H. T. Cory, "Water Supply of the San Francisco-Oakland Metropolitan District with Discussion by F. T. Robson, et al." *Transactions* 50 (1916): 43.

74. Anson Barstow, "Annual Message" (Oakland: 1902), p. 7.

75. Warren Olney, "Address to Citizens of Oakland," 14 February 1903, in *Oakland Miscellany*, vol. 1:19, Bancroft Library, Berkeley.

76. Marks, "From the Beginning," pp. 5–6, 13–15.

77. Chamberlain et al., "Municipal Ownership of Water and Available Sources of Supply," pp. 2, 5; Warren Olney, "Address to Citizens of Oakland," 14 February 1903, in *Oakland Miscellany*, vol. 1:19, Bancroft Library, Berkeley.

78. Anson Barstow, "Annual Message" (Oakland: 1902), p. 7.

79. Berkeley City Club, Committee on Water Supply, "Report," in *Berkeley Civic Bulletin* 1:12 (1913): 131, Water Resources Center Archive, Berkeley.

2. Beyond Municipal Boundaries

1. Suburbs welcomed annexation because it gave them access to Boston's lavish public services. Like other central cities, Boston sought additional territory for prestige and because annexation increased access to natural resources, tax revenues, and influence in the state legislature. Annexation also appealed to Boston's leaders because it allowed a single administration to coordinate services for a large territory. The combination of these benefits made annexation so attractive that Boston actually rejected a number of cooperative projects that might have reduced suburbs' need for Boston's services.

2. Los Angeles not only annexed extensive suburban areas, but also appropriated vast and distant water resources. The Owens Valley Aqueduct has been thoroughly studied as an example of Los Angeles' political manipulation and urban power in rural areas. See, for example, Donald Worster, *Rivers of Empire: Water, Aridity and the Growth of the American West* (New York: Pantheon Books, 1985); Norris Hundley, Jr., *The Great Thirst: Californians and Water, 1770s–1990s* (Berkeley: University of Califor-

nia Press, 1992); William L. Kahrl, *Water and Power: The Conflict over Los Angeles' Water Supply in the Owens Valley* (Los Angeles: University of California Press, 1982).

3. Boston Board of Health, 1875, p. 5. Boston's death rate rose from 23.5 per 1,000 in 1871 to 30.4 per 1,000 in 1872 before falling slightly to 28.5 per 1,000 in 1873. In contrast, European cities at the same time had mortality rates that remained closer to 21 per 1,000.

4. "Public Health," *Boston Post,* 13 November 1874.

5. Boston city councilors debated the creation of a comprehensive park system extensively during November and December, 1873. As early as 1869, a public meeting on the topic attracted many Bostonians who favored construction of at least a large central facility in the city. See Boston City Council, Joint Special Committee . . . to Consider . . . a Public Park, *Report and Accompanying Statements and Communications Relating to a Public Park for the City of Boston,* Boston: 1869, pp. 3–5; *City Council Proceedings* 1873, pp. 462–64, 482, 549–54. See Christine M. Rosen, *The Limits of Power: Great Fires and the Process of City Growth in America* (Cambridge: Cambridge University Press, 1986), for a detailed account of the Boston fire of 1872 and of subsequent efforts to improve transportation and city planning.

6. Boston Water Board, *Boston Water Works: Additional Supply from the Sudbury River* (Boston: 1882), pp. 7, 9; Desmond Fitzgerald, *History of the Boston Water Works from 1868 to 1876* (Boston: 1876), p. 87.

7. Fitzgerald, *History of the Boston Water Works,* p. 38.

8. Howe, *Quabbin,* p. 10; "Inquiry into the Best Mode of Supplying the City of Boston with Water for Domestic Purposes" (Boston: 1845), p. 23, in *Water Reports 1835–1845,* pamphlet collection, Massachusetts Water Resources Authority Library, Charlestown, Massachusetts. John B. Jervis, the most prominent of the Cochituate engineers, is better known for his part in designing the Croton Aqueduct for New York City.

9. A significant portion of this increase in per capita water use can be attributed to industrial rather than domestic consumption. See Joel A. Tarr, James McCurley, and Terry F. Yosie, "The Development and Impact of Urban Wastewater Technology: Changing Concepts of Water Quality Control, 1850–1930," in Martin V. Melosi, ed., *Pollution and Reform in American Cities, 1870–1930* (Austin: University of Texas Press, 1980), p. 62. Although running water did transform overall water use patterns, many Americans had experimented with running water while still relying solely on wells or cisterns. For details, see Maureen Ogle's *All the Modern Conveniences: American Household Plumbing, 1840–1890* (Baltimore: Johns Hopkins University Press, 1996).

10. Roxbury and Dorchester brought 47,031 new residents to Boston; annexation of Charlestown, Brighton, and West Roxbury added another 54,400 to the census. These suburbs accounted for 15 to 19 percent of Boston's total population. United States Census, *The Seventh Census of the United States: 1850* (Washington, D.C.: 1853); Department of the Interior, United States Census, *Ninth Census, vol. 1: Statistics of the Population of the United States* (Washington, D.C.: 1872); Department of the Interior, United States Census, *Statistics of the Population of the United States . . . Tenth Census* (Washington, D.C.: 1883).

11. "Records of the City of Boston Mayor and Aldermen, 1848," p. 82, manuscript record books, *Uncatalogued Boston City Records Collection,* Boston Public Library, Boston, Massachusetts.

12. Chapter 177, *General Court Acts 1872.*

13. Boston Water Board, *Boston Water Works,* p. 7; "Report of the Commissioners Appointed to Investigate the Great Fire of Boston" (Boston: 1873), p. viii.

14. Boston Water Board, *Boston Water Works,* pp. 28–31; for reactions to the Great Fire, see Fitzgerald, *History of Boston Waterworks,* p. 37. See also for a comparison of Boston's Great Fire to those in other cities.

15. "Remarks of Alonzo Warren of Ward 12 ... on Additional Supply of Water," 14 May 1874, no. 50 in *Documents of the City of Boston for the Year 1874* (Boston: 1875), pp. 3–6, 10, 13, 14; "Remarks of Francis H. Peabody, of Ward 9, in the Common Council, May 14, 1874 on an Additional Supply of Water," no. 51 *Documents of the City of Boston for the Year 1874* (Boston: 1875), pp. 3, 4, 8.

16. "The Water Question," *Boston Sunday Herald,* 15 November 1874.

17. "An Engineer ..." *Boston Globe,* 16 May 1877.

18. "Water Supply," *Boston Post,* 13 November 1874.

19. Chapter 177, *General Court Acts 1872;* Chapter 158, *General Court Acts 1875.* The General Court protected the water claims of Framingham, Ashland, Southborough, Hudson, and Westborough, but granted all other water in the river north of Framingham to Boston. Subsequent legislation prevented Hopkinton from taking water out of the river.

20. Fitzgerald, *History of the Boston Water Works,* p. 9.

21. Boston Board of Health, *Ordinances Prescribing Rules and Regulations Relative to Nuisances, Sources of Filth, and Causes of Sickness within the City of Boston* (Boston: 1833).

22. Tarr, McCurley, and Yosie, "The Development and Impact of Urban Wastewater Technology," pp. 61–63.

23. Boston City Council, Committee on Sewers, *Report* (Boston: 1874), pp. 3–5; Boston Sewer Department, *Annual Report,* 1868, pp. 7–8.

24. From 1868 to 1875, the Boston sewer department installed more than 36 miles of pipe in the annexed territories and nearly 22 miles of pipe in Boston proper. Construction continued at these rates throughout the 1870s and 1880s. By way of comparison, Boston built just over 2 miles of new pipe per year in the three years before annexation began. The invention of precast cement pipe made this rapid increase in city sewage lines possible. Boston Sewer Department, *Annual Reports,* 1865 through 1889.

25. Boston City Council, *Report of the Committee on the Petition of David Sears and Others in Respect to Drainage of the Back Bay,* no. 14 in *Documents of the City of Boston for the Year 1850* (Boston: 1850), pp. 4–6, 8, 12.

26. Boston Sewer Department, *Annual Report of the Superintendent of Sewers 1860* (Boston: 1861), pp. 6–8; Boston Sewer Department, *Annual Report 1861* (Boston: 1862), pp. 12–13; Boston Sewer Department, *Annual Report 1867* (Boston: 1868), pp. 6–7.

27. *City Council Proceedings 1876,* 174–75. These businesses included the Jordan Marsh Company (a retailer still active in New England), the George C. Richardson, William Claflin, and May companies, and 720 other petition signers.

28. Boston Sewer Department, *Annual Report 1868* (Boston: 1869), p. 8; Boston Sewer Department, *Annual Report 1869* (Boston: 1870), p. 10; *City Council Proceedings 1878,* pp. 21, 32, 446.

29. In 1796, John Lowell, Increase Sumner, Thomas Williams, John Reed, and Thomas Williams, Jr., formed the corporation to dig the canal. Until they completed the project, the only channel to Roxbury passed between Boston and South Boston. This route was essentially impassable at low tide. See Thomas C. Simonds, *History of South Boston: Formerly Dorchester Neck, New Ward XII of the City of Boston* (Boston: David Clapp, 1857), p. 70.

30. Boston Board of Health, "Communication from the Board of Health upon the Necessity of Improved Sewerage in the Roxbury Canal, Stony Brook and Muddy Brook," no. 112 in *Documents of the City of Boston for the Year 1874* , pp. 3–5.

31. In 1878, A. D. Williams, from the working-class South End, and David W. Cheever, a physician, both submitted petitions demanding that the city council clean the canal. *City Council Proceedings 1878*, pp. 21, 32, 446.

32. The health officials' involvement in sewerage reforms contributed to the creation of permanent health agencies in both the state and the city. The Massachusetts State Board of Health, founded in 1867, was one of the first permanent health agencies in the nation. The State Board wielded significant influence in Boston's sewerage and water supply debates over the next thirty years. Beginning in 1799, Boston, like most other American communities, had created and disbanded health departments in response to specific epidemics or other crises. Boston replaced its temporary health organizations with a permanent Superintendent of Health in 1853, and with a fully independent city Board of Health in 1872. See Barbara Gutman Rosenkrantz, *Public Health and the State: Changing Views in Massachusetts, 1842–1936* (Cambridge, Mass.: Harvard University Press, 1972), p. 1; and Census Office, "Social Statistics of Cities," pp. 122–23.

33. Boston Board of Health, *A Communication from the Board of Health to the Committee on Improved Sewage* (Boston: 1876), pp. 3, 6.

34. Massachusetts Medical Commission, *The Sanitary Condition of Boston. The Report of a Medical Commission to the Boston Board of Health* (Boston: 1875), pp. 171–72.

35. Massachusetts Medical Commission, *The Sanitary Condition of Boston*, pp. 82–83, 101.

36. Massachusetts Medical Commission, *The Sanitary Condition of Boston*, p. 172.

37. *City Council Proceedings 1864*, pp. 222, 225.

38. Boston Committee on Sewers, "Report on the Present System of Sewerage in the City of Boston," no. 94 in *Documents of the City of Boston for the Year 1873*, p. 8.

39. E. S. Chesbrough, Moses Lane, and Charles F. Folsom, *The Sewerage of Boston, a Report by a Commission* (Boston: Rockwell & Churchill, 1876), pp. 1, 7.

40. For more information on Chesbrough's accomplishments, see Louis P. Cain's *The Search for an Optimum Sanitation Jurisdiction: The Metropolitan Sanitary District of Greater Chicago, A Case Study*, Essays in Public Works History, no. 10 (1980), and "Raising and Watering a City: Ellis Sylvester Chesbrough and Chicago's First Sanitation System," in Judith Walzer Leavitt and Ronald L. Numbers, eds., *Sickness and Health in America: Readings in the History of Medicine and Public Health* (Madison: University of Wisconsin Press, 1985), pp. 439–50.

41. *The National Cyclopaedia of American Biography* (Ann Arbor: University Microfilms, 1967), 4: 377; 6: 34, 35.

42. Chesbrough, Lane, and Folsom, *Sewerage of Boston*, pp. 1–7, 9–10, 12.

43. *Discharge of Sewage*, p. 19; *City Council Proceedings 1877*, p. 561.

44. Chesbrough, Lane, and Folsom, *Sewerage of Boston*, p. 18, 23–24; Eliot C. Clarke, *Main Drainage Works of the City of Boston* (Boston: 1885), pp. 18–19.

45. *City Council Proceedings 1876*, p. 358; Eliot C. Clarke, *Main Drainage Works of the City of Boston* (Boston: 1885), pp. 18–19. The Commission of 1875 projected a budget of $3,746,500 for the interceptors south of the Charles. The northern interceptors would have cost an additional $2,804,564, according to Chesbrough, Lane, and Folsom.

46. *City Council Proceedings 1877*, pp. 570–71, 574–76.

47. *City Council Proceedings 1877*, pp. 570–71, 574–76.

48. *City Council Proceedings 1877*, pp. 560, 571, 576; George W. Fuller, "Paper for the International Engineering Congress, St. Louis, Missouri, October, 3–8, 1904," compiled in *Papers Relating to Sewerage*, vol. 1, pp. 1–4; Tarr, McCurley, and Yosie, "Urban Wastewater Technology," pp. 69–70. A letter published in the *Boston Journal* alluded to the British experience to draw similar conclusions about the hazards of ocean disposal. See Festina Lente, "The Plan of Sewage," *Boston Journal*, 2 August 1877.

49. *City Council Proceedings 1877*, pp. 559–61.

50. *City Council Proceedings 1877*, pp. 567, 579. During this period, the aldermen were elected at large. Consequently, common council votes give a somewhat more accurate indication of the geographical distribution of support for any given issue.

51. *City Council Proceedings 1877*, p. 569.

52. Of the four primarily poor working-class neighborhoods, only the North End failed to cast any common council votes for the sewers; all three of its representatives abstained. Aside from the North End, the wards with the highest immigrant and working-class populations—in South Boston, the South End, South Cove, and parts of Roxbury and Dorchester—strongly supported the project. Two South End representatives from ward 11 did abstain; but as ward 11 included part of the Back Bay, these abstentions probably reflected continued enthusiasm for the Back Bay park. The Back Bay's other wards, 18 and 22, voted four in favor and one opposed to the project. The southern suburbs and Beacon Hill, areas of considerable wealth, voted strongly in favor of the project. See *City Council Proceedings 1877*, p. 567.

53. See Robin Einhorn, *Property Rules: Political Economy in Chicago, 1833–1872* (Chicago: University of Chicago Press, 1991), pp. 6–8.

54. In 1868, the Boston Sewer Department triumphantly reported the complete enclosure of Stony Brook, "to the general satisfaction of the abutters as well as the city." This marked the permanent conversion of the creek into one of Boston's sewer mains. See Boston Sewer Department, *Annual Report 1868*, p. 8.

55. John H. Hooper, "Sewage in Mystic River," *Medford Historical Register* 23 (1920), p. 47; "Report of the Mystic Water Board on an Order of the City Council Relating to Sewerage, " in vol. 3, *Documents of the City of Boston for the Year 1874* , pp. 5–6; *City Council Proceedings 1873*, p. 443; *City Council Proceedings 1876*, p. 52.

56. Medford, *Report of the Board of Selectmen on the Subject of Sewerage* (Medford, Mass.: Chronicle Press, 1873), pp. 3–5; Hooper, "Sewage in Mystic River," pp. 45–46. Except in direct quotes—where Mystic Lake and Mystic Pond are used interchangably—I have used Mystic Lake to denote Charlestown's old reservoir. I have referred to the lower lake, which was never used as a water supply, as Mystic Lower Pond to distinguish it from the reservoir.

57. *City Council Proceedings 1874,* pp. 290, 316; *City Council Proceedings 1876,* p. 51; *City Council Proceedings 1875,* p. 426; *City Council Proceedings 1874,* pp. 617, 623; *City Council Proceedings 1875,* pp. 426, 483; Legislative Packet for Chapter 220, *General Court Acts 1875,* Massachusetts Archives, Boston; Chapter 202, *General Court Acts 1875.*

58. Chapter 453, *General Court Acts 1874; City Council Proceedings 1874,* p. 617.

59. *City Council Proceedings 1874,* p. 617.

60. *City Council Proceedings 1876,* pp. 52–53.

61. The Mystic Water Board projected that a sewer large enough for domestic as well as industrial wastes would cost $600,000. Early estimates for the Main Drain exceeded $6,550,000. See "Report of the Mystic Water Board on an Order of the City Council Relating to Sewerage," no. 51 of *Documents of the City of Boston for the Year 1874* (Boston: 1875), pp. 3–4; Timothy T. Sawyer for the Mystic Water Board, in *City Council Proceedings 1874,* pp. 611–12; and *Remonstrance of George B. Emerson and Other Taxpayers of Boston, against the Adoption of the System of Sewerage Proposed in Report No. 3* (Boston: 1876), p. 3.

62. One councilor insisted that the city should not spend water or tax revenues "at the bidding of a few men owning summer residences in that vicinity." See *City Council Proceedings 1876,* p. 53.

63. *City Council Proceedings 1875,* p. 461.

64. Hooper, "Sewage in Mystic River," p. 49.

65. *House Journal 1875,* p. 375; "Remonstrance of the Selectmen of the Town of Arlington and 130 Others . . . ," 15 April 1875, in General Court Legislative Packet for Chapter 202, *General Court Acts 1875,* Massachusetts State Archives, Boston.

66. Hooper, "Sewage in Mystic River," p. 46; *City Council Proceedings 1876,* pp. 51–52.

67. *City Council Proceedings 1876,* p. 54.

68. Chapter 202, *General Court Acts 1875; Senate Journal 1881,* p. 321; *House Journal 1881,* p. 23.

69. *House Journal 1881,* pp. 399, 433; Massachusetts General Court, *Documents of the House of Representatives for the Year 1881* (Boston: 1882), p. 329; "Committee on Public Health, Report of Minority," in Legislative Packet for Chapter 303, *General Court Acts 1881,* Massachusetts State Archives.

70. Chapter 303, *General Court Acts 1881.*

71. "Petition of Daniel W. Lawrence and Others of Medford for Such Legislation as Will Protect the Purity of the Water in Mystic Lower Pond . . ." 10 January 1881, Legislative Packet for Chapter 303, *General Court Acts 1881,* Massachusetts State Archives, Boston.

72. The fact that East Bay health officials played a declining role in public works did not mean that Californians did not fear epidemics. On the contrary, health officials throughout the state spent much of the 1910s battling bubonic plague. Plague was endemic in China early in the century. When, in 1900, the first confirmed case was diagnosed in San Francisco's Chinatown, the United States Health Service imposed a quarantine on all Chinese immigrants. The Chinese community saw the quarantine not as a public health measure, but as a continuation of decades of anti-Chinese politics in California. When the disease reappeared after a three-year respite, health departments sought to control the plague by exterminating squirrels in public parks. By 1910, these efforts had eliminated the danger of a full-blown epidemic,

but had not eradicated the disease. Bubonic plague is still endemic in rodents throughout California and parts of the western United States. Vernon B. Link, *A History of Plague in the United States of America,* Public Health Monographs No. 26 (Washington, D.C.: 1955), pp. 1–23, 27–42.

73. Oakland grew from 16.6 to 60.1 square miles as a result of these annexations. Judith V. May, "Progressives and the Poor: An Analytic History of Oakland," presented at Public Administration and Neighborhood Control conference, Boulder, Colorado, May 6–8, 1970, pp. 28–31, 51.

74. Marsden Manson and C. E. Grunsky, *Report on the Plans and Estimates for the Proposed Main Sewers East of Lake Merritt as Submitted to the City Council of Oakland* (San Francisco: 1893), pp. 4–7, Bancroft Library, Berkeley.

75. "Mayor and Moorehead in Wordy Battle," 18 August 1924, *San Francisco Chronicle,* and "Elmhurst's One Gunman Routs Gang," 16 November 1925, *Oakland Tribune,* both in *Oakland Sewers 1919–1930* file, Oakland History Room, Oakland Public Library, Oakland, California.

76. "City Seeks Truce with Elmhurst," September 1925, *Oakland Tribune,* and "Elmhurst Army Wins War Waged on Storm Sewer; City Surrenders," 17 November 1925, *Oakland Tribune,* both in Oakland History Room, *Oakland Sewers 1919–1930* file, Oakland Public Library.

77. "Baccus' Plan on Sewer Hit by Piedmont," 18 February 1927, *Oakland Tribune,* in Oakland History Room, *Oakland Sewers 1919–1930* file, Oakland Public Library.

78. John R. Davie, "Mayor's Message" (Oakland: 1929), p. 45.

79. "Engineer Tells City's Sewer System Need," 7 June 1933, *San Francisco Chronicle,* Oakland History Room, *Oakland Sewers 12/9/30-7/18/41* file, Oakland Public Library.

80. Committee of East Bay City Engineers, "Preliminary Report upon Sewage Disposal for the East Bay Cities of Alameda, Albany, Berkeley, El Cerrito, Emeryville, Oakland, Piedmont, San Leandro and Richmond" (Oakland: 1938), pp. 8, 44, Bancroft Library; Charles Gilman Hyde, Harold Farnsworth Gray, and A. M. Rawn, *Report upon the Collection, Treatment and Disposal of Sewage and Industrial Wastes of the East Bay Cities, California* (np: 1941), pp. 9, 54–55.

81. Philip E. Harroun, *Report to the Commission of the East Bay Cities on the Water Supply for the Cities of Oakland, Berkeley, Alameda and Richmond,* p. 8.

82. Harroun, *Report to the Commission of the East Bay Cities,* 9; Arthur P. Davis, George W. Goethals, and William Mulholland, *Additional Water Supply of the East Bay Municipal Utility District: A Report to the Board of Directors* (Oakland, 1924) p. 6.

83. Unsigned letter to Mayor and City Council of Berkeley, 3 May 1921, Louis Bartlett Papers, box 13, *East Bay Water Co.* file, Bancroft Library, Berkeley.

84. J. B. Spears to Bartlett, 2 May 1921, Bartlett Papers, box 13, *East Bay Water Co.* file, Bancroft Library, Berkeley.

85. "Water or Bondage? Which Is It? Business Men's Reasons for Opposing Bonds," (1905), p. 5, in *Oakland Miscellany,* vol. 4:5, Bancroft Library, Berkeley.

86. Arthur L. Adams, "How Shall Oakland Secure Public Ownership of Water Works?" (np: 1904), pp. 8, 10.

87. Adams, "How Shall Oakland Secure Public Ownership of Water Works?" p. 4. $5.7 million in 1904 is approximately $87 million in 1992 dollars.

88. The fact that, in 1900 and again in 1906, Bay Cities tried to sell Alameda water

rights to San Francisco makes the offer to Oakland seem all the more speculative. See Marsden Manson, "Preface: Outline of the History of the Water Supply of the City of San Francisco," in C. E. Grunsky and Marsden Manson, *Report on the Water Supply of San Francisco, California, 1900 to 1908, Inclusive* (San Francisco: 1909), pp. 10–11.

89. R. H. Chamberlain, John L. Howard, Warren Olney, Sol Kahn, and James P. Taylor, *Municipal Ownership of Water and Available Sources of Supply for Oakland, California* (Oakland: Enquirer, 1903), p. 13, in *Oakland Miscellany*, vol. 4:4, Bancroft Library, Berkeley; Adams, "How Shall Oakland Secure Public Ownership of Water Works?" pp. 5–6.

90. "Will Mayor Olney Stop His Water War?" *Oakland Tribune*, May 1904, in John Edmund McElroy Scrapbooks, vol. 2, Bancroft Library, Berkeley.

91. See Donald Pisani, *From the Family Farm to Agribusiness: The Irrigation Crusade in California and the West, 1850–1931* (Berkeley: University of California Press, 1984), for a good explanation of the development of the agencies that governed water rights in California.

92. According to John T. Cumbler, "The Early Making of an Environmental Consciousness: Fish, Fisheries Commissions, and the Connecticut River," *Environmental History Review* 15 (1991), common law protected public fishing rights in ponds of more than 10 acres. This protection did not survive industrial development. Mill dams constructed along the Concord River in the mid-nineteenth century flooded pastures traditionally used by farmers. The conflict between agriculture and industry came to a head between 1860 and 1862 and was finally resolved in industry's favor when the General Court protected the rights of dam builders to impound water despite damages to pastures. The legislature, like the judiciary before it, demonstrated with this decision that it wished to avoid disrupting the development of industry along the river and in the state generally. See Brian Donahue, "'Damned at Both Ends and Cursed in the Middle:' The 'Flowage' of the Concord River Meadows, 1798–1862," *Environmental Review* 13 (1989), pp. 48–49, 57.

93. "Will Mayor Olney Stop His Water War?" *Oakland Tribune*, May 1904, in John Edmund McElroy Scrapbooks, vol. 2, Bancroft Library, Berkeley; "Water or Bondage?" pp. 4–6.

94. According to Annmarie Hauck Walsh, states began imposing debt limits on municipal administrations in the mid-1800s. At the time, cities had invested heavily in railroads and other private utilities. When those companies failed, mixed public-private ownership and municipal overinvestment threw many American cities into bankruptcy. See Walsh, *The Public's Business: The Politics and Practices of Government Corporations* (Cambridge, Mass.: MIT Press, 1978), pp. 19–23.

95. "The Special Water Committee of City Council Submits Its Report," 19 January 1904, unnamed Oakland paper in John Edmund McElroy Scrapbooks, vol. 2, Bancroft Library, Berkeley.

96. "Will Mayor Olney Stop His Water War?" *Oakland Tribune*, May 1904, in John Edmund McElroy Scrapbooks, vol. 2, Bancroft Library, Berkeley.

97. Cory, "Water Supply of the San Francisco–Oakland Metropolitan District," p. 43; John Wesley Noble, *Its Name Was M.U.D.* (Oakland: East Bay Municipal Utility District, 1970), p. 11.

98. Marsden Manson, "Preface: Outline of the History of the Water Supply of the City of San Francisco," in Grunsky and Manson, *Reports on the Water Supply of San Francisco,* pp. 5, 7.

99. San Francisco Bureau of Engineering (M. M. O'Shaughnessey), *Hetch Hetchy Water Supply* (San Francisco: 1925), p. 6. A number of books trace the Hetch Hetchy controversy. See, in particular, Roderick Nash, *Wilderness and the American Mind* (New Haven, Conn.: Yale University Press, 1973); and Warren D. Hanson, *A History of the Municipal Water Department and Hetch Hetchy System* (City and County of San Francisco: 1985).

100. J. H. Dockweiler, *Report on Sources of Water Supply: East Region of San Francisco Bay* (San Francisco: 1912), p. 9.

101. Mel Scott, *San Francisco Bay Area: A Metropolis in Perspective* (Berkeley: University of California Press, 1985), pp. 133–34.

102. Scott, *San Francisco Bay Area,* pp. 134, 141.

103. "Company is Blamed for Water Shortage," *Oakland Post,* 12 September 1918, in *Clippings 1916–1920,* p. 38, East Bay Municipal Utility District Records Office, Oakland, California; Harroun, *Report to the Commission of the East Bay Cities,* pp. 8–9.

104. "Supreme Court on Monopoly Valuations," *San Francisco Call,* 12 July 1909, in John D. Galloway Papers, 82–1, Water Resources Center Archive, Berkeley; "Water Co. to Ask Higher Water Rate," *San Francisco Enquirer,* 18 September 1918, in *Clippings 1916–1920,* p. 30, East Bay Municipal Utility District Records Office; "Peoples Water Properties Claimed Overvalued $5,100,000," *Oakland Tribune,* 6 February 1917; "Water Company Expert Testifies at Rate Hearing," *San Francisco Chronicle,* 2 February 1917, both in *Clippings 1916–1920,* 31, East Bay Municipal Utility District Records Office.

105. "Peoples Water Co. Will Use San Pablo Creek Project as Basis for Claim that Richmond Consumers Should Pay More for Water," *Richmond Record Herald,* 16 December 1916 in *Clippings 1916–1920,* p. 3, East Bay Municipal Utility District Records Office.

106. "Another Water Combine," *Oakland Press* 2 February 1921(?), in *Clippings 1916–1920,* p. 35, East Bay Municipal Utility District Records Office. The fact that Leland Stanford, one of the Big Four who built the transcontinental Central Pacific, controlled a major interest in San Francisco's Spring Valley Water Company lent a certain credibility to accusations of an East Bay railroad-water company connection.

107. William Deverell, "Building an Octopus: Railroad and Society in Late Nineteenth Century." Ph.D. diss., Princeton University, 1989, pp. 120, 131–32. According to Donald Pisani, the individuals who monopolized Kern County's water supplies directed attention away from their power by blaming Southern Pacific for the failure of small farms in the Kern County area. See Pisani, *From the Family Farm to Agribusiness,* p. 202. See also Spencer C. Olin, *California's Prodigal Sons: Hiram Johnson and the Progressives, 1911–1917* (Berkeley: University of California Press, 1968), pp. 2, 45–50; Mansel G. Blackford, *Politics of Business in California, 1890–1920* (Columbus, Ohio: Ohio State University Press, 1977), pp. 78–80.

108. "Platform Adopted by the Sacramento Convention" (np, 1902), in John Edmund McElroy Scrapbooks, vol. 1, Bancroft Library, Berkeley.

109. Walter Clark, "Public Ownership the Only Solution," *Pacific Municipalities* 37:1 (1923): 3–8.

110. Johnson's campaign strategy consisted almost exclusively of berating the railroad and blaming the Southern Pacific "octopus" for nearly all of the state's economic and political woes. See Olin, *California's Prodigal Sons.*

111. Deverell, "Building an Octopus," pp. 375–412; Mansel G. Blackford, *Politics of Business in California, 1890–1920* (Columbus, Ohio: Ohio State University Press, 1977), p. 85.

112. Michael Paul Rogin and John L. Shover, *Political Change in California: Critical Elections and Social Movements, 1890–1966* (Westport, Conn.: Greenwood, 1970), pp. 46–49.

113. Jon C. Teaford's *Unheralded Triumph: City Government in America, 1870–1900* (Baltimore: Johns Hopkins University Press, 1984) criticizes traditional urban historiography for focusing narrowly on the conflicts between bosses and reformers, and accepting the rhetoric of this conflict as an accurate representation of urban conditions. He argues that a better measure of urban government comes from close examination of government structures, policies, and accomplishments. After conducting such a study, Teaford concludes that, given the challenges facing late-nineteenth-century cities—including simultaneous demands for low taxes and extensive services—urban leaders indeed triumphed by meeting most demands and maintaining a flexible, balanced political structure.

114. "Water or Bondage?" pp. 1–3.

3. Boston: Regionalism in the Gilded Age

1. Nancy Burns, *Formation of American Local Governments: Private Values in Public Institutions* (New York: Oxford University Press, 1994), pp. 14, 46, 48–53; Winston W. Crouch and Dean E. McHenry, *California Government: Politics and Administration* (Berkeley: University of California Press, 1954), p. 194; Annmarie Hanck Walsh, *The Public's Business: The Politics and Practices of Government Corporations* (Cambridge, Mass.: MIT Press, 1978), pp. 19–23.

2. *Fifth Annual Report of the State Board of Health, Lunacy and Charity of Massachusetts* (Boston: 1884), p. lxvii; Massachusetts State Board of Health, *Examinations . . . of the Water Supplies and Inland Waters of Massachusetts, 1887–1890. Part 1 of Report on Water Supply and Sewerage* (Boston: 1890), p. 36. These industries, the Board of Health noted, were located too far up small tributaries to be connected to the Mystic Valley Sewer.

3. John H. Hooper, "Sewage in Mystic River," *Medford Historical Register* 23 (1920): 45–53. *City Council Proceedings 1882,* pp. 10, 54.

4. Medford, *Report of the Board of Selectmen on the Subject of Sewerage* (Medford, Mass.: Chronicle Press, 1873), pp. 3–5; Hooper, "Sewage in Mystic River," pp. 45–46.

5. In 1889, the State Board of Health reported that restrictions on waste disposal in streams, combined with inaction on the regional sewerage system by the General Court, delayed local sewage projects in Medford, Woburn, and other Mystic towns. See Massachusetts State Board of Health, *Annual Report for 1889,* pp. 5, 6.

6. John D. Long, "Inaugural Address of his Excellency John D. Long," in *General Court Acts 1880,* p. 680.

7. Chapter 62, *General Court Acts 1881*; *Senate Journal 1881*, pp. 361, 373, 402.

8. *Fifth Annual Report of the State Board of Health, Lunacy and Charity of Massachusetts* (Boston: 1884), p. lxvii; *Third Annual Report of the State Board of Health, Lunacy and Charity of Massachusetts* (Boston: 1882), p. lxv.

9. In 1879, the Massachusetts State Board of Health was combined with several other state agencies to create the State Board of Health, Lunacy and Charity. Many members of the original board saw this as a disastrous project that condemned health policy to the vagaries of partisan politics. The health agency regained its independent status in 1886. Neither of these reorganizations affected the state health officials' stance on regionalism. See Barbara G. Rosenkrantz, *Public Health and the State: Changing Views in Massachusetts, 1842–1936* (Cambridge, Mass.: Harvard University Press, 1986).

10. *Third Annual Report of the Massachusetts State Board of Health, Lunacy and Charity* (Boston: 1882), p. lxiv.

11. William Ripley Nichols and George Derby, p. 106.

12. William Ripley Nichols, *Sewerage; Sewage; the Pollution of Streams; the Water Supply of Towns* (Boston: 1873), as quoted in *Boston Board of Health 1875*, pp. 112–15. The lack of legislative means to halt water contamination before it began may have contributed to legislation passed in 1878 that did allow the State Board of Health to seek court injunctions in situations like this one.

13. Chapter 228, *General Court Acts 1875*, "An Act to Preserve the Purity of the Water of Lake Cochituate," passed 19 May 1875.

14. Massachusetts Metropolitan Drainage Commission, "Report on the Mystic and Charles River Valleys Drainage," no. 35 in *Documents of the City of Boston for the Year 1882* (Boston: 1882), p. 12.

15. Charles W. Folsom, "Scrapbook on State Drainage," p. 62, Boston Public Library, Boston, Massachusetts.

16. Folsom, "Scrapbook on State Drainage," p. 62.

17. E. S. Chesbrough, Moses Lane, and Charles Follen Folsom, *The Sewerage of Boston: A Report by a Commission* (Boston: 1876), pp. 10–12. Chesbrough, Lane, and Folsom identified some twenty sewers from Cambridge and Somerville alone that emptied into the Charles River.

18. Massachusetts State Board of Health, *Annual Report for 1878*, p. 348.

19. Edward R. Cogswell, "Report on the Sanitary Condition of Cambridge," in Massachusetts State Board of Health, *Annual Report for 1878*, p. 347.

20. *City Council Proceedings 1882*, p. 223.

21. *Fifth Annual Report of the Massachusetts State Board of Health, Lunacy and Charity*, (Boston: 1884), p. lxviii.

22. Massachusetts Metropolitan Sewerage Commissioners, "Main Drainage Works of Boston and Its Metropolitan Sewerage District" (Boston: 1899), p. 11.

23. Chapter 439, *General Court Acts 1889*.

24. "Minority Report, House Committee on Drainage," House Document 410, *House Documents 1889*, pp. 1–6.

25. Chapter 424, *Massachusetts Acts and Resolves 1899* (27 May 1899).

26. Because the Neponset was an important industrial stream, it was exempt from most early state prohibitions on sewage disposal. By 1900, sewage in the Neponset

and nearby marshes appeared to threaten local communities with malaria and other diseases. Local communities, much as they feared disease, rejected a State Board of Health proposal to improve river drainage by removing industrial dams. That, they felt, would cripple Neponset industries; the southern drain received support from these communities because it cleaned the river without interfering with industrial production. See Massachusetts General Court, Committee of Ways and Means, "Hearing on a Bill Relative to the Protection of the Public Health in the Valley of the Neponset River," 9 May, 1900, pp. 10, 12–15, 18–25, including comments by E. C. Bumpus on behalf of Tileston & Hollingsworth and the Baker Chocolate Company.

27. Adelaide M. Cromwell, *The Other Brahmins: Boston's Black Upper Class, 1750–1950* (Fayetteville: University of Arkansas Press, 1994), pp. 48–49. In the 1882 campaign, Albert E. Pillsbury secured the following endorsement, citing Pillsbury's family credentials, from Boston's black newspaper: "He is the nephew of Parker Pillsbury, who was one of the anti-slavery sons of thunder. His uncle, Hon. Gilbert Pillsbury, was the first Mayor of Charleston, S. C., after the war. He and his remarkable wife were the idols of the blacks in Charleston. . . . He is himself a true friend of our race, a liberal, able and public-spirited Republican. . . . *Every colored man in the 6th district should vote for him*" [emphasis in original]. ("A. E. Pillsbury" unsigned clipping, 1882, in A. E. Lindbury [Pillsbury], "Republican Party Politics in Boston, 1873–82," General Scrapbook 65, Huntington Library, San Marino, California.)

28. "Irish Democrats in Massachusetts," *Boston Pilot,* 5 May 1877.

29. "General Butler and the Irish Voter," *Boston Pilot,* 14 September 1878.

30. In 1878, these conflicts so divided Democrats, that Irish Democrats split the party over this exclusion, voting for Benjamin Butler instead of Josiah Abbott, the candidate chosen by party leadership."An Outrage by Party Managers," *Boston Pilot* 21 September 1878; "General Butler and the Irish Voter," *Boston Pilot,* 14 September 1878; Thomas H. O'Connor, *The Boston Irish: A Political History* (Boston: Northeastern University Press, 1995), pp. 109–10. For a more complete discussion of ethnic politics in Boston, see also Oscar Handlin, *Boston's Immigrants: A Study in Acculturation,* rev. ed. (Cambridge, Mass.: Belknap Press, 1979); Ronald P. Formisano and Constance K. Burns, eds., *Boston 1700–1980: The Evolution of Urban Politics* (Westport, Conn.: Greenwood, 1984).

31. "Brakes on City Extravagance," *Boston Daily Globe,* 12 June 1877.

32 "The Clashing of Systems," *Boston Journal* 24 May 1875. For their part, Irish leaders tried to deflect insinuations of corruption by urging Irish voters to place honesty and merit above national identification. In some cases, voters were urged to cast their ballots for Republican Know-Nothings, despite their anti-immigrant invective, rather than elect "a scalawag Irishman." They reasoned that "[e]very loud-mouthed Irish demagogue elected to office this year will leave a stain in that office which our children will have to wipe out with hard work and bitter humiliation." See "The Coming Elections, Who Shall Represent Us?" *Boston Pilot,* 1 November 1873.

33. "Brakes on City Extravagance," *Boston Daily Globe,* 12 June 1877.

34. "City Commissions," *Boston Daily Journal,* 30 June 1874.

35. "Want Suffolk County. Boston Republicans Already Pulling Wires for Annexation," *Boston Post,* 13 September 1895.

36. "Want Suffolk County. Boston Republicans Already Pulling Wires for Annexation," *Boston Post,* 13 September 1895.

37. "Free Cities," *Boston Post,* 14 April 1881.

38. Patrick Maguire, as quoted in O'Connor, *The Boston Irish,* p. 149.

39. As Barbara Rosenkrantz explains in *Public Health and the State: Changing Views in Massachusetts, 1842–1936* (Cambridge: Harvard University Press, 1972), the creation of the State Board of Health, Lunacy and Charity in 1879 and reorganization of the State Board of Health in 1886 reflected partisan rivalry, efforts to limit state spending, the fragmentation of the Republican Party by the Mugwumps, and the influence of "nonpartisan" reformers.

40. *Boston Evening Transcript,* 6 November 1878; *Boston Evening Transcript,* 9 November 1881; *Boston Evening Transcript,* 9 November 1887.

41. Massachusetts State Board of Health, "Report . . . upon the Discharge of Sewage into Boston Harbor," (Boston: 1900), p. 21; Nathan Matthews, Opening Statement, "Third Hearing," 7 November 1904, vol. 1, *Cities of Malden, Medford and Melrose v. Commonwealth of Massachusetts,* Commonwealth of Massachusetts Superior Court, Middlesex County, p. 38, Boston Public Library.

42. Massachusetts State Board of Health, *Report . . . upon the Discharge of Sewage into Boston Harbor,* p. 22; Massachusetts State Board of Health, *Report . . . upon a Metropolitan Water Supply* (Boston: 1895), p. xiii; Massachusetts General Court, *Legislative Hearings as to the Nashua River Water Supply before the Committees on Metropolitan Affairs and Water Supply,* 25 February to 29 April 1895, pp. 53–54, 69, State Library, Boston. (Hereafter cited as *Nashua Hearings.*) Engineers designed water systems with capacities large enough to supply cities during inevitable dry spells. This minimum expected flow into reservoirs and wells was called the "safe yield" of the water system. Using more water than the safe yield—called "overdraft"—left city water systems without enough water to carry them through even short droughts.

43. *City Council Proceedings 1894,* pp. 999, 1057; Massachusetts State Board of Health, *Annual Report for 1895,* p. 87.

44. Edward S. Wood, "The Water Supply of Cambridge," in Massachusetts State Board of Health, *Annual Report for 1879,* pp. 70–75.

45. "Shawsheen" is now the accepted spelling of the river's name. During the 1880s, "Shawshine" and other forms were also used. I have used the modern spelling here, but retained other forms as they appear in quotations or the names of primary documents.

46. "Evidence and Arguments on Petitions of Cambridge and Boston for Leave to Take Water from the Shawshine River, before the Committee on Public Health of the Massachusetts Legislature" (Boston: Franklin Press, 1882), pp. 5–9, in *Boston and Cambridge Water Supply* , Widener Library, Harvard University (hereafter cited as "Shawshine 1882"); *City Council Proceedings 1882,* p. 555.

47. "Shawshine 1882," 3.

48. Desmond Fitzgerald, *History of the Boston Water Works,* p. 87; "Shawshine 1882," p. 3.

49. *City Council Proceedings 1882,* p. 48.

50. *City Council Proceedings 1882,* p. 54.

51. "Shawshine 1882," pp. 194–95; "Shawshine 1882," pt. 4, p. 66.

52. In a series of disputes over the Concord River in the mid-1860s, Massachusetts courts gave mill owners greater opportunities to construct water power dams despite

the damage this industrial development did to farming and fishing rights along the stream. See Brian Donahue, "'Damned at Both Ends and Cursed in the Middle:' The 'Flowage' of the Concord River Meadows, 1798–1862," *Environmental Review* 13 (1989), pp. 48–49, 57.

53. "Shawshine 1882," pp. 3–4, 194–95.

54. "Shawshine 1882," pp. 11, 23; *Senate Journal 1882*, p. 42. Cambridge abandoned its campaign for the Shawsheen in 1882. Two years later, however, the city did get legislative approval for a different water supply project, this one tapping Stony Brook in Waltham and Weston. See Chapter 256, *General Court Acts 1884*.

55. Boston Water Board, *Special Report . . . to the City Council, September 6, 1883* (Boston: 1883), pp. 1–2, 8.

56. Donahue, "'Damned at Both Ends and Cursed in the Middle," pp. 48–49, 57.

57. *City Council Proceedings 1894*, p. 999; Boston, Water Supply Department, "Annual Report for 1892," no. 39 in *Documents of the City of Boston for 1893*, p. 11.

58. Nathan Matthews, Jr., *The City Government of Boston* (Boston: 1895), p. 65; Chapter 459, *General Court Acts of 1893*.

59. "Report of the State Board of Health upon a Metropolitan Water Supply," in Massachusetts State Board of Health, *Annual Report for 1894* (Boston: 1894), p. cxvii.

60. George C. Whipple, *State Sanitation: A Review of the Work of the Massachusetts State Board of Health* (New York: Arno Press, 1877), pp. 179–80; William L. Kahrl, *Water and Power: The Conflict over Los Angeles' Water Supply in the Owens Valley* (Berkeley: University of California Press, 1982), p. 150.

61. "Report of the State Board of Health upon a Metropolitan Water Supply," pp. cxviii, cxix.

62. Massachusetts General Court, *Nashua Hearings*, pp. 14–15; "Report of the State Board of Health upon a Metropolitan Water Supply," pp. cxxii–iii. Although Stearns did not mention it in his report, extensive industrial use of Lake Winnepesaukee and the Merrimack River watershed posed an additional barrier to development of the lake, as Theodore Steinberg describes in *Nature Incorporated: Industrialization and the Waters of New England* (Cambridge: Cambridge University Press, 1991).

63. "Report of the State Board of Health upon a Metropolitan Water Supply," pp. cxxiv–ix. In the 1920s, no real consensus existed that the metropolitan area needed additional water supplies. The Quabbin was built without extensive reexamination of the central assumptions made regarding Wachusett in 1895. Because the regional water network already existed when the Quabbin was approved, I have not examined it here. For a complete discussion of the Quabbin, see Fern L. Nesson, *Great Waters: A History of Boston's Water Supply* (Hanover, N.H.: University Press of New England, 1983) or Donald W. Howe, *Quabbin, the Lost Valley* (Quabbin Book House, Mass.: Ware, 1951).

64. "Opposed to a Greater Boston," *Boston Daily Globe*, 7 November 1895, in *Boston Water Supply: Clippings from Boston Newspapers, 1895–1904*, Rare Books Room, Boston Public Library; *Nashua Hearings*, pt. 4, p. 50.

65. *Nashua Hearings*, pt. 4, p. 48; *Nashua Hearings*, pt 5, pp. 64–66.

66. *Nashua Hearings*, pt. 4, pp. 50, 54; *Nashua Hearings*, pt. 5, p. 46.

67. Massachusetts Metropolitan Drainage Commissioners, *Report on the Mystic and Charles River Valleys Drainage*, p. 11; *City Council Proceedings 1890*, p. 901.

68. "Report of the State Board of Health upon a Metropolitan Water Supply," 87–102; *City Council Proceedings 1895*, p. 826; *House Journal 1895*, p. 1200.

69. *Nashua Hearings*, pt. 5, pp. 14, 22–23, 52, 64–66; *Nashua Hearings*, 178.

70. Arlington's position in this debate is quite ironic. Even though the community declined to join the regional water district, it was initially among those lobbying for compulsory membership. *Nashua Hearings*, pt. 5, p. 48; *Nashua Hearings*, p. 173; *House Journal 1895*, p. 1200.

71. *City Council Proceedings 1896*, p. 53; *City Council Proceedings 1895*, p. 826.

72. *City Council Proceedings 1896*, p. 52. This order passed in January 1896.

73. John Koren, *Boston, 1822 to 1922: The Story of Its Government and Principle Activities During One Hundred Years* (Boston: 1923), p. 98.

74. Koren, *Boston, 1822 to 1922*, p. 99; Frederick P. Stearns, "Information Relating to the Claims of the Town of Clinton," 23 February 1901, pp. 1–5, State House Library, Boston; Howe, *Quabbin*, p. 23;*Nashua Hearings*, p. 51. Wachusett Reservoir claimed 1417 acres of Clinton's total 4645 acres, and 33 Clinton residences. In Boylston, the metropolitan project displaced 303 of the town's 770 inhabitants and 46 of its 155 dwellings. The reservoir also eliminated lands worth $165,000 from the town's total $429,000 tax rolls. West Boylston lost $557,730 of its $933,945 valuation and 157 of its 480 dwellings, housing 1305 of West Boylston's 3019 residents.

75. "Water Grabbing," *Clinton Daily Item*, 1 February 1896.

76. *Worcester Gazette* as quoted in "Boston's Water Waste," *Clinton Daily Item*, 28 March 1895.

77. *Nashua Hearings*, p. 193; "West Boylston's Argument in Regard to the Metropolitan Water Supply" (Worcester: O. B. Wood, 1895); "Report of the State Board of Health upon a Metropolitan Water Supply," p. xix; Stearns, "Information Relating to the Claims of the Town of Clinton," pp. 1–5.

78. *Nashua Hearings*, p. 192; *Nashua Hearings*, pt. 5, pp. 54–55, 58, 61.

79. *Nashua Hearings*, p. 192.

80. "West Boylston Views," *Clinton Daily Item*, 31 March 1896.

81. "Sawyers Mills: History of a Busy Village on the Nashua to be Closed by the Removal of the Mills," *Clinton Daily Item*, 4 March 1899; "Boylston, April 9," in *C. A. Cook Scrapbook*, vol. 2, 1899; "West Boylston, Jan. 29," in *C. A. Cook Scrapbook*, vol. 1, "1897–1899," Beaman Memorial Library, West Boylston, Massachusetts.

82. "Central Mass. Road. Recent Surveys for New Location Through Clinton," *Clinton Daily Item*, 2 October 1901.

83. "An Ejectment. Two Families on Boylston Street Removed from their Homes," *Clinton Daily Item*, 6 August 1898.

84. "West Boylston Views," *Clinton Daily Item*, 31 March 1896.

85. "Clinton Woman: The Victim of an Engineer Who Came to Work on the Metropolitan Reservoir," *Clinton Daily Item*, 6 September 1901.

86. "Speculators after Land," *Clinton Daily Item*, 16 February 1895.

87. "Laborers on Strike. Work Stops at the Big Dam at West Boylston," *Worcester Telegram* 1903, in *C. A. Cook Scrapbook*, vol. 4, "December 1902–May 1906."

88. Section 31, "An Act to Provide for a Metropolitan Water Supply," Chapter 488, *General Court Acts of 1895*; Jill Lepore, "Area Chosen as Reservoir Site," *Clinton Daily Item*, 5 July 1989. Jill Lepore has documented some of the social and economic reper-

cussions of the reservoir in Clinton. See also Jill Lepore, "Immigrant Workers Lived under Primitive Conditions," *Clinton Daily Item,* 6 July 1989.

89. "Clinton Laborers Protest," *Clinton Daily Item,* 13 August 1898.

90. "Laborers on Strike. Work Stops at the Big Dam at West Boylston," *Worcester Telegram,* 1903, in *C. A. Cook Scrapbook,* vol. 4, "December 1902–May 1906"; "Italians Strike: the Gang of Men Employed by Busch Bros., on the New Road, Threaten Trouble," *Clinton Daily Item,* 1 September 1899.

91. "Lorato Demarzio is Stabbed to Death! . . . Result of Quarrel over Payment for Three Drinks of Cider," *Worcester Telegram,* 1900, in *C. A. Cook Scrapbook* vol. 2.

92. "A Big Steal: Two Italians Take a Whole Grocery Outfit from Clinton to Worcester," *Clinton Daily Item,* 6 October 1899.

93. "Sale of Liquor," *Clinton Daily Item,* 15 July 1899; "A Liquor Hunt," *Clinton Daily Item,* 24 July 1899.

94. "Clinton Laborers Protest," *Clinton Daily Item,* 13 August 1898.

95. "Investigation," *Clinton Daily Item,* 26 January 1900; "The Investigation," *Clinton Daily Item,* 1 February 1900.

96. "Investigation," *Clinton Daily Item,* 26 January 1900.

97. Charles F. Choate, Jr., "Brief and Argument in Behalf of Clinton Business Men's Association before a Special Joint Committee to Investigate Violations of Law on Wachusett Reservoir, " (np, nd). State Library, Boston.

98. See Robert Gottlieb and Margaret Fitzsimmons, *Thirst for Growth: Water Agencies as Hidden Governments in California* (Tucson: University of Arizona Press, 1991) for a discussion of this pattern in California's water institutions.

99. "Report by the Commission Appointed to Investigate the Subject of Annexation of Certain Neighboring Cities and Towns to the City of Boston" (Boston: 1874), pp. 6–7; Alfred D. Chandler, "Annexation of Brookline to Boston: Opening Argument for the Town of Brookline before the Committee on Towns of the Massachusetts Legislature," 11 March 1880 (Brookline: 1880), pp. 7–12. In 1895, Democratic city councillors rejected a Boston water department reorganization as a blatant attempt by Mayor Curtis to replace Democrats with Republican appointees. See "May Be a Fight," *Boston Daily Globe,* 15 November 1895; "Mr. Quincy Writes," *Boston Journal,* 25 November 1895; "Mayor Curtis: Charges against Administration," *Boston Daily Globe,* 29 November 1895.

100. *Nashua Hearings,* 1. The Metropolitan Affairs Committee in 1895 included four Republican senators and nine Republican assemblymen. The only Democrats, two assemblymen, came from Boston. The Water Supply Committee included nine Republicans and two Democrats. The Democrats represented Cambridge and Boston.

101. In 1878, the General Court passed "An Act Relative to the Pollution of Rivers, Streams and Ponds Used as Sources of Water Supply," Chapter 183, *General Court Acts 1878.* This legislation prohibited the discharge of sewage within 20 miles of a water supply intake, and gave power of enforcement to the State Board of Health. The act included two exemptions that severely limited its ability to protect drinking water: it specifically recognized existing rights of drainage, and it exempted the state's most heavily industrialized rivers, the Merrimack, Connecticut, and Concord. Joel Tarr, in "The Separate Versus Combined Sewer Problems: A Case Study in Urban Tech-

nology Design Choice," *Journal of Urban History* 5 (1979) pp. 329–33, finds that treatment of drinking water quickly supplanted pollution prevention as the best means to ensure clean drinking water. This decision turned attention away from sewage treatment, and permitted the continued contamination of the nation's rivers. Massachusetts' 1878 act testifies to the pressures ranged against strict water pollution control legislation that might have interfered with industrial development in New England.

4. The East Bay: Regionalism in the Progressive Era

1. The Contra Costa Water Company (CCWC), organized in 1866, was the first major water utility in the East Bay. The Oakland Water Company was incorporated in 1893, and joined CCWC in a major rate war from 1895 to 1899. In 1899, the two water companies merged, retaining the Contra Costa Water Company name. In 1900, CCWC acquired Alameda Artesian Water Company, then supplying Berkeley. In 1906, CCWC was reorganized as Peoples Water Company, serving most of Alameda County. The following year, Peoples Water acquired the Richmond Water Company. Now the water company provided services in both Alameda and Contra Costa counties. In 1916, Peoples Water reorganized as the East Bay Water Company; the monolithic company now owned networks and reservoirs developed by a total of 18 independent companies. For complete details of these mergers, see John Wesley Noble, *Its Name Was M.U.D.* (Oakland: East Bay Municipal Utility District, 1970).

2. See Note 107, Chapter Two.

3. Michael Paul Rogin and John L. Shover, *Political Change in California: Critical Elections and Social Movements, 1890–1966* (Westport, Conn.: Greenwood, 1970), pp. 46–49.

4. "Another Water Combine," *Oakland Press,* 2 February 1921(?), in *Clippings 1916–1920,* p. 35, East Bay Municipal Utility District Records Office, Oakland, California.

5. Walter Clark, "Public Ownership the Only Solution," *Pacific Municipalities* 37:1 (1923): 3–8.

6. Spencer C. Olin, Jr., *California Politics, 1846–1920: The Emerging Corporate State* (San Francisco: Boyd & Fraser, 1981).

7. Philip E. Harroun, *Report to the Commission of the East Bay Cities on the Water Supply for the Cities of Oakland, Berkeley, Alameda and Richmond* (np: 1920), pp. 12,13; Noble, *Its Name Was M.U.D.,* pp. 4, 172.

8. In 1997, EBMUD provided water to 1.2 million customers in a 325 square mile area. East Bay regional sewerage serves 600,000 people in 83 square miles. In contrast, the Los Angeles Department of Water and Power serves 3.6 million customers in a territory covering 465 square miles. The Los Angeles Department of Water and Power should not be confused with the Metropolitan Water District of Southern California. The latter agency delivers water to a 5,200-square-mile territory and approximately 16 million people, including those tapping the Los Angeles Department of Water and Power system. San Francisco's Water Department distributes water to 770,000 San Franciscans and 1.6 million suburban customers in four counties.

9. S. M. Marks, "From the Beginning: An Outline of the History of the Water Supply for the East Bay District from 1854–1910" (np, 1922?), p. 10, in File 388, East Bay Municipal Utility District Archives.

10. Charles Gilman Hyde, "An Extensive Outbreak of Gastro-Enteritis in the City of Alameda, California, from Polluted Water Consequent upon Defective Well Casings," 25 October 1944, in Hyde Papers, p. 15, Water Resources Center Archive, Berkeley, California. The outbreak struck 38.2 percent of Alamedans. Other communities were spared because they did not receive water from the contaminated Fitchburg wells.

11. J. H. Dockweiler, *General Information Regarding Proposed Metropolitan Municipal Water District, Alameda County, California,* np, pp. 4–5, Bancroft Library, Berkeley.

12. Annmarie Hauck Walsh, *The Public's Business: The Politics and Practices of Government Corporations* (Cambridge, Mass.: MIT Press, 1978), p. 22. California Constitution, art. 4, §37 (1849), sets municipal tax and debt limits. The section reads: "It shall be the duty of the legislature to provide for the organization of cities and incorporated villages, and to restrict their power of taxation, assessment, borrowing money, contracting debts and loaning their credit, so as to prevent abuses in assessments and in contracting of debts by such municipal corporations."

13. "Water Project of the City is Legalized," *San Francisco Call,* 26 December 1911, p. 5. See California Senate Bill 1028 (1911).

14. Max Thelan, "The Public Utilities Act and Its Relation to Municipalities," *Pacific Municipalities* 26(2) (1912), p. 50; John M. Eshleman, "The Regulation of Public Utilities," *Pacific Municipalities* 26(7) (1912), pp. 349–50.

15. Pisani, *From the Family Farm to Agribusiness,* pp. 358–64.

16. Eshleman, "The Regulation of Public Utilities," pp. 345–50.

17. C. H. McNary, "Competition, Regulation or Municipal Ownership," *Pacific Municipalities* 26(5) (1912), pp. 228–32. McNary was formerly an engineer for Los Angeles Edison.

18. C. H. McNary, "Competition, Regulation or Municipal Ownership," *Pacific Municipalities* 26(5) (1912), pp. 228–32; Blackford, *Politics of Business,* pp. 78–95, passim.

19. D. J. Hall, "State Regulation of Municipally Owned Utilities," *Pacific Municipalities* 37(6) (1923), pp. 223–24. Mansel G. Blackford argues in *The Politics of Business* that distributing franchises provided city officials with too many opportunities for graft.

20. Dockweiler, "General Information Regarding Proposed Metropolitan Municipal Water District," p. 6.

21. Noble, *It's Name Was M.U.D.,* p. 14. Oakland's final vote tally was 6,478 for the water district and 9,142 against.

22. William Dingee to Frank C. Havens, 6 January 1911, in Frank C. Havens Papers, Box 1, Bancroft Library, Berkeley.

23. "Peoples Water Co. Will Use San Pablo Creek Project as Basis for Claim that Richmond Consumers Should Pay More for Water," *Richmond Record Herald,* 16 December 1916, in *Clippings 1916–1920,* p. 3.

24. "Richmond Women Favor Preserving the District," *Richmond News,* 25 January 1917, in *Clippings 1916–1920,* p. 21. The *Richmond News* polled prominent women, including the owner of the Fairview Hotel, and members of the Richmond Club.

25. "Remove Water Tax Menace, Plan. People Flock to Sign Petition," *Richmond Independent,* 22 December 1916, in *Clippings 1916–1920,* p. 5.

26. "Union Men and Women Urged to Vote Tuesday," *Richmond Independent,* 25 January 1917, p. 27; "City Council Under Charter Can Conduct Municipal Water," *Richmond Daily Independent,* 18 January 1917, p. 11, both in *Clippings 1916–1920.*

27. "Local Public Service Corporation Trying to Thwart Will of Voters Here and Bind City to Cast Away Opportunity to Own Water System," *Richmond Record Herald,* 1 December 1916; "Richmond Women Favor Preserving the District," *Richmond News,* 25 January 1917, both in *Clippings 1916–1920,* pp. 1, 21.

28. "Richmond Women Favor Preserving the District," *Richmond News,* 25 January 1917, in *Clippings 1916–1920,* p. 21.

29. "Water District Is Disincorporated," *Richmond News,* 31 January 1917, in *Clippings 1916–1920,* p. 29.

30. "Public Utilities League to Beverly J. Hodgehead, 8 July 1915, in Galloway Papers, 82-1, Water Resources Center Archive; "Barrow Tells Vast Scope of Water Project," *Oakland Tribune,* 10 December 1924, *EBMUD Bond Election 1924* file, Oakland History Room, Oakland Public Library.

31. "4 Cities in Water Session," *San Francisco Bulletin,* 3 December 1918; "Measure Attacks Magoon Scheme," *Oakland Bulletin* 29 July 1918, both in *Clippings 1916–1920,* 10.

32. Carl D. Thompson to Bartlett, 3 November 1931, Bartlett Papers, box 4, Bancroft Library, Berkeley.

33. "Attorneys Debate Water Problem," np, 1920, p. 8, and "Davie Proposes New Commission," *Oakland Daily Post,* 29 July 1918, p. 10, both in *Clippings 1916–1920*; "Davie and Magoon in Final Clash: Statements End Utilities Fight," *Oakland Tribune,* 24 August 1918, and "Magoon Makes Reply to Davie," *Oakland Enquirer,* 31 July 1918, *EBMUD 7/18–1/31/26* file, Oakland History Room, Oakland Public Library.

34. "Mayor Plans Commission to Control Public Utilities," *Oakland Enquirer,* 29 July 1918, *Clippings 1916–1920,* p. 39.

35. "Mayor Plans Commission to Control Public Utilities," *Oakland Enquirer,* 29 July 1918, *Clippings 1916–1920,* p. 39.

36. John L. Davie was also extremely proud of his efforts to break transportation monopolies in Oakland. See John L. Davie, *His Honor, the Buckaroo: The Autobiography of John L. Davie* (Reno, Nev.: Jack Herzberg, 1988), pp. 13–16; and William Deverell, *Railroad Crossing: Californians and the Railroad, 1850–1910* (Berkeley: University of California Press, 1994), pp. 156–58.

37. The final tally was three to two against the district. Oakland rejected the district more strongly than any other community, while Berkeley voters approved the measure three to one. See "Utility Plan Loses; Charter Changes Win," *Oakland Tribune,* 28 August 1918, *EBMUD 7/18–1/31/26* file, Oakland History Room, Oakland Public Library; "People's Water Co. and All Subsidiaries Will Default," *San Francisco Chronicle,* 27 June 1914; Noble, *It's Name Was M.U.D.,* pp. 12–14; "Company Is Blamed for Water Shortage," *Oakland Post,* 12 September 1918, in *Clippings 1916–1920,* p. 38.

38. Pisani, *From the Family Farm to Agribusiness,* p. 58. Drought visited the state in 1863–1864, 1898–1899, and 1929–1931, and more recently in 1976–1977 and 1986–1992. Population growth has made each successive drought worse. In 1992, the state had so few reserves that, for the first time, it had to reduce guaranteed water allocations to farmers throughout the state.

39. Arthur Powell Davis, George W. Goethals, and William Mulholland, *Additional*

Water Supply of East Bay Municipal Utility District: A Report to the Board of Directors (Oakland: 1924), pp. 5–6. Saltwater intrusion would have permanently contaminated the Alvarado aquifer, rendering it unfit for human consumption. Overdraft in the South Bay Area caused land around San Jose to subside by 4 feet from 1910 to 1933, and by an additional 9 feet since then. Santa Clara Valley Water District, "Average Depth to Water, Land Subsidence, Population" (San Jose: Santa Clara Valley Water District Public Information Office, 1992).

40. Louis Bartlett, *Memoirs*, interview by Corinne L. Gilb, University of California Regional Cultural History Program (Berkeley, 1957), p. 98.

41. Bartlett, *Memoirs*, p. 99.

42. "Increased Rates May Help Public," *Oakland Enquirer*, 14 October 1919, *Clippings 1916–1920*, p. 40.

43. Philip E. Harroun, *Report to the Commission of the East Bay Cities on the Water Supply for the Cities of Oakland, Berkeley, Alameda and Richmond* (np, 1920), pp. 56–58, Bancroft Library, Berkeley.

44. Harroun, *Report to the Commission of the East Bay Cities*, pp. 56–58.

45. Rockwell D. Hunt, ed., *California and Californians*, vol. 3 (San Francisco: Lewis, 1926), pp. 470–71; "A Statement of Principles by Anna L. Saylor," campaign literature 1924, California State Library; *Eminent Californians*, 2d ed. (Palo Alto, Calif.: C. W. Taylor, 1956), 43; "Utility District Is Decided Upon," *Oakland Tribune*, 16 February 1921, in *EBMUD 7/18–1/31/26* file, Oakland History Room, Oakland Public Library. Both Progressives, Saylor and Spence shared an association with Clement C. Young, the Progressive leader who was elected as governor in 1929. Saylor was elected to the seat that Young abandoned in 1918. Spence served Young as a secretary in 1927. William Locke of Alameda, Frank V. Cornish of Berkeley, and Leon Gray of Oakland helped draft the Municipal Utility District Act.

46. "Attorneys Debate Water Problem," np 1920, in *Clippings 1916–1920*, p. 8; Bartlett, *Memoirs*, p. 99.

47. "Berkeley Mayor Opposes Public Utility Measure," *Oakland Tribune*, 17 February 1921, in *EBMUD 7/18–1/31/26* file, Oakland History Room, Oakland Public Library.

48. California, *Laws, Statutes, etc.*, "Organic Act of the East Bay Municipal Utility District Act" (1935), §6, Bancroft Library.

49. Mary S. Gibson, *A Record of Twenty-five Years of the California Federation of Women's Clubs, 1900–1928*, pp. 133–34. These goals were listed in the Federation's 1914 legislative agenda. Bartlett to L. E. Blochman, 17 March 1920, in *Bartlett Papers*, box 13; Bartlett to Ira A. Morris, 17 March 1920, in *Bartlett Papers*, box 13; "Civic Organizations Appoint Committees to Water Commission," *Berkeley Times* (nd), in *Clippings 1916–1920*, p. 44.

50. Bartlett, *Memoirs*, pp. 106–8.

51. "Plan Campaign for Utility District," *Berkeley Gazette*, 26 October 1920; "As the Post Interprets the News," *Oakland Post*, 30 April 1920.

52. Clearly, after EBMUD began delivering water to the East Bay, industrial developers and promoters recognized this advantage of regional water. By 1932, the Chamber of Commerce touted the large number of industrial sites served with EBMUD water that were located in unincorporated territories. Companies in the East Bay

and elsewhere have found these areas, free of land-use restrictions and municipal property taxes, particularly attractive for industrial development. See Oakland Chamber of Commerce, Industrial Water Committee, "Report on Cost of Water as Related to Industrial Development in the East Bay Region" (Oakland: Oakland Chamber of Commerce, 1932), Water Resources Center Archive.

53. Gibson, *A Record of 25 Years of the California Federation of Women's Clubs, 1900–1925*, vol. 1 (California Federation of Women's Clubs, 1927), pp. 133–34; "Plan Campaign for Utility Districts," *Berkeley Gazette*, 26 October 1920; "Labor Acts in Mokelumne Project," *East Bay Labor Journal*, 21 November 1925; "Attacking Our Water Project," *East Bay Labor Journal*, 17 January 1925; Walter Clark, "Public Ownership the Only Solution," *Pacific Municipalities* 37(1) (1923), pp. 3–4.

54. Walter Clark, "Public Ownership the Only Solution," *Pacific Municipalities* 37(1) (1923), pp. 3–4; Olin, *California's Prodigal Sons*, pp. 46–49; Rogin and Shover, *Political Change in California*, pp. 38, 51, 75.

55. Michael Kazin, *Barons of Labor: The San Francisco Building Trades and Union Power in the Progressive Era* (Urbana: University of Illinois, 1987). For more information on the relationship between Progressives and labor, see Olin, *California's Prodigal Sons*, pp. 46–49; Rogin and Shover in *Political Change in California*, pp. 38, 51, 75.

56. Walter Clark advocated public ownership as a means to end labor agitation in "Public Ownership the Only Solution," *Pacific Municipalities* 37(1) (1923), pp. 3–8; George Pardee, long associated with California's Progressive Party, was mayor of Oakland during the 1894 Pullman Strike. Despite tremendous public support for the strikers, Pardee called in the National Guard and portrayed strikers as lawless revolutionaries. See Beth Bagwell, *Oakland: The Story of the City* (Novato, Calif.: Presidio Press, 1982), pp. 80–81, and Deverell, *Railroad Crossing*, pp. 77–79, 233.

57. Archaeological excavations of nineteenth- and twentieth-century water company construction camps have revealed artifacts indicating that Chinese laborers helped build the East Bay's reservoirs and aqueducts before EBMUD. East Bay Municipal Utility District, lobby display 1991, Oakland, California.

58. Bartlett, *Memoirs*, pp. 99–100; "Officials Balk at Public Utilities District Drive," *Oakland Tribune*, 20 January 1923.

59. "Mayor Plans Commission to Control Public Utilities," *Oakland Enquirer*, 29 July 1918, in *Clippings 1916–1920*, p. 39.

60. "Mayor Plans Commission to Control Public Utilities," *Oakland Enquirer*, 29 July 1918; "Mayor Heard on Utility Project," *Oakland Tribune*, 12 August 1918; "Davie Proposes New Commission," *Oakland Daily Post*, 29 July 1918; "Magoon Makes Reply to Davie," *Oakland Enquirer*, 31 July 1918.

61. "Davie in Favor of Consolidation," *San Francisco Chronicle*, 3 January 1916; City and County Government Association, "Some of the Benefits that Would Accrue under a City and County Charter" (Oakland, 1918) in *Oakland, California: Politics and Government* pamphlet collection, Bancroft Library, Berkeley.

62. "Davie in Favor of Consolidation," *San Francisco Chronicle*, 3 January 1916; Anti-Division League, "County Division Means Increased Taxes," in *Oakland, California: Politics and Government* pamphlet collection, Bancroft Library, Berkeley.

63. "Davie Proposes New Commission" *Oakland Daily Post*, 29 July 1918, in *Clippings 1916–1920*, p. 10.

64. Bartlett, *Memoirs,* pp. 104–5; "Berkeley Rejects Utility District," *Oakland Daily Post,* 15 November 1920, p. 12; "Charter Amendments," editorial in *Berkeley Gazette,* 1 November 1920, p. 12, all in *Clippings 1916–1920.* In 1920, Bartlett had used this same strategy, in the form of a Berkeley-Richmond water project, to coax Richmond to join the Water Commission. See "Mayor Louis Bartlett and Members of Water Commission Ask Richmond to Join City of Berkeley in Municipal Water Scheme," *Richmond News,* 2 March 1920; "Water Proposition Takes another Turn," np, p. 39; "Water District Election Is Called Off," np, p. 39, all in *Clippings 1916–1920.*

65. "Eastbay Cities Join Forces to Acquire Service," *Oakland Tribune,* 11 June 1920, in *Clippings 1916–1920,* p. 46; Taxpayers Association of California, "Expenses and Outlays of the California State Government for . . . 1902–1917" (np, 1918), p. 16.

66. Louis Bartlett, "Subtle Influences against Water Plan: Coterie Fears Dividend Loss," *Oakland Post-Enquirer,* 23 April 1923; "Surf Sirens in 'Skimpy' Suits Smash 'Semblage," *Oakland Daily Post,* 13 March 1919; Franklin Hichborn, *California Politics, 1891–1939* (Haynes Foundation, nd), pp. 1852–53, 2542. Hichborn asserted that these organizations were "financed, officered and directed by the public service corporations." In his book, *The California Progressives* (Berkeley: University of California Press, 1951), George Mowry explains that the California Taxpayers Association, a statewide organization representing the state's twelve largest corporations, sponsored a campaign of corporation-focused legislation to defend the prerogatives of private utilities. Taxpayers' leagues frequently arose to oppose Progressive measures that threatened to increase public spending. In the 1930s, the California Taxpayers' Association and the California Development Association, financed by and acting on behalf of "public service corporations," overshadowed Progressive influence in California state politics.

67. "Higher Costs Halts Upper River Fight," *Lodi Sentinel,* 13 May 1926, p. 1; "Pardee Fires Hot Shot at Millar Plan," *Lodi Sentinel,* 4 February 1926, p. 12.

68. "East Bay Utilities at Odds with East Bay Co.," *Lodi News,* 3 April 1925, in *Clippings 1916–1920,* p. 11.

69. Charles Gillman Hyde, "The Sanitary Control of Community Water Supplies as Illustrated by Work of the Department of Sanitation of the Peoples Water Company," p. 116, in *Charles Gillman Hyde Papers,* Water Resources Center Archive; Hyde, "Notes on a Water Supply for the East Bay Cities Utilizing the Flood or High-Stage Waters of the Lower Sacramento and San Rivers and a Comparison with the Proposed Mokelumne Project" (Berkeley, 1924), pp. 5–8, in *Charles Gillman Hyde Papers* box 24, Water Resources Center Archive.

70. "Water Fight Concluded by Pardee," *Oakland Tribune,* 7 May 1925, and "Mokelumne Bid Call Due Next Week," *Oakland Tribune,* 27 May 1925, both in *EBMUD 7/18–1/31/26* file, Oakland History Room, Oakland Public Library.

71. "East Bay Utilities at Odds with East Bay Co.," *Lodi News,* 3 April 1925, in *Clippings 1916–1920,* p. 11.

72. Piedmont and Richmond joined the district a few months later; Castro Valley and Lafayette followed in 1931. East Bay Municipal Utility District, *The Story of Water: A Brief History of the East Bay Municipal Utility District and a Description of the Source, Transmission, Treatment and Distribution of the Water Supply to the Cities within Its Boundaries* (Oakland: East Bay Municipal Utility District, 1931), p. 4; "Testimony of Eugene

Sullivan, President of Blue Lakes Water and Power Company, before the U.S. Senate Committee on the Public Lands, 23 July 1913," pp. 1–2, in "Mokelumne River Project; Hetch Hetchy; SB 2610 file," East Bay Municipal Utility District Records Office. Of the 47,335 voters casting ballots, 29,914 voted in favor of EBMUD and 17,421 voted against.

73. The fire began as a grass fire in Wild Cat Canyon. High winds and wooden shingle roofs allowed the fire to spread quickly and consume nearly 600 buildings and inflict over $10 million in property damage. Low water pressure was listed as one of the major hindrances to effective fire fighting. See National Board of Fire Underwriters, "Report on the Berkeley, California Conflagration of September 17, 1923," in Miscellaneous Publications Regarding the Berkeley, California Fire of September 17, 1923, Green Library, Stanford University, Palo Alto, California.

74. "Board Seeks Data on Water," *Oakland Post,* 7 March 1924, in *EBMUD 7/18–1/31/26* file, Oakland History Room, Oakland Public Library.

75. "Directors Not Pledged to Eel River," *Oakland Tribune,* 16 February 1924, in *EBMUD 7/18–1/31/26* file, Oakland History Room, Oakland Public Library; Bartlett, *Memoirs,* p. 111.

76. Rolph to Bartlett, 24 June 1924, quoted in Bartlett, "Public Ownership in and around San Francisco" (np, 1931?), p. 14, *Bartlett Papers,* box 4; "Water Deal is Deadlocked," *Oakland Tribune,* 11 August 1924, in *EBMUD 7/18–1/31/26* file, Oakland History Room, Oakland Public Library.

77. Bartlett, "Public Ownership in and around San Francisco," (np, 1931?), p. 16, *Bartlett Papers,* box 4, Bancroft Library.

78. "Water Scheme Is Attacked in Questionnaire," *Lodi Sentinel,* 23 October 1924, p. 4; "Makes Light of Attempt to Save Mokelumne," *Lodi Sentinel,* 13 October 1925, p. 1.

79. Arthur P. Davis, "Sources of Supply for the East Bay Water District," (1912), pp. 11, 14, in *Sources of Supply; Arthur P. Davis; 1924* file, East Bay Municipal Utility District Records Office, Oakland; "Alameda Offers Water Grant to Utility," *Oakland Post,* 12 July 1923, and "Campbell Denies Eel River Plan," *Oakland Tribune,* 15 February 1924, both in *EBMUD 7/18–1/31/26* file, Oakland History Room, Oakland Public Library.

80. "M'Cloud River Offered as Water Supply," *San Francisco Chronicle,* 23 April 1925; Davis, Goethals, and Mulholland, *Additional Water Supply,* p. 16; East Bay Municipal Utility District, *The Story of Water,* p. 5.

81. Arthur McEvoy, *The Fishermen's Problem: Ecology and Law in the California Fisheries, 1850–1980* (Cambridge: Cambridge University Press, 1986), pp. 83–84, 87. Mining eliminated salmon runs in many Mother Lode rivers from 1850 to 1890. Fish returned to some of these streams after hydraulic mining ended in the 1890s, but never in the numbers that they had before 1852.

82. Stephen E. Kieffer, "How Much-Sought Reservoir Sites Were Discovered," *Stockton Record,* 8 November 1924.

83. Davis, Goethals, and Mulholland, *Additional Water Supply* , p. 63.

84. In 1913, the California legislature approved a bill that established the priority of municipal water development over all private projects. See Pisani, *From the Family Farm to Agribusiness,* p. 366.

85. "Double Legal Tangle Holds Up Issue of $39,000,000 Bonds Voted for Mokelumne Project," *Lodi Sentinel*, 3 February 1925, p. 1.

86. "Broken Faith Is Charged to the East Bay," *Stockton Record*, 16 October 1925, p. 2; "U. S. Grants Water Right to Preston," *Lodi Sentinel*, 17 October 1925.

87. *Joint Hearing before the Division of Water Rights, Department of Public Works, State of California and the Federal Power Commission*, 11–12 September 1925, reporter's transcript, p. 999, East Bay Municipal Utility District Records Office.

88. Pardee to O. C. Gould, 15 May 1926, in Pamphlet Box of Materials on EBMUD, Bancroft Library; *Joint Hearing before the Division of Water Rights*, pp. 960–61, 982–83, 1007–10.

89. Unsigned, "Pertinent Data Relative to the Cost of Lands Sought to Be Condemned in the Case of East Bay Municipal Utility District vs. Kieffer, et al.," nd, *Bernard A. Etcheverry Papers*, file 84, Water Resources Center Archive; Franklin Hichborn, "California Politics, 1891–1939" (Haynes Foundation, nd), pp. 8–9, Stanford University Library; Pardee to O. C. Gould, 15 May 1926, *Pamphlet Box of Materials on EBMUD*, Bancroft Library, Berkeley. According to one report, Kieffer demanded over $15 million for 8 acres of land along the best route between Calaveras Cement Mill and the Lancha Plana dam site. See "Kieffer Price for Calaveras Lands Scored," *Lodi Sentinel*, 1 March 1927.

90. "Private Interests Renew Fight to Gain Control of Mokelumne," *Lodi Sentinel*, 4 May 1926, p. 1; "Kieffer Is Held Finder of Dam Site, *Lodi Sentinel*, 4 June 1927, p. 1.

91. "Water Suit Is Not Surprise, States Hyatt," *Lodi Sentinel*, 6 May 1926.

92. Charles W. Slack, Edgar T. Zook, Arthur C. Huston, and F. J. Solinsky, "Opening Brief for J. W. Preston, Jr.," in *Joint Hearing before the Division of Water Rights*, pp. 23, 30, 40.

93. *Joint Hearing before the Division of Water Rights*, pp. 10, 12, 1583–86, 1752–53; "Mokelumne Foes Fail in Fight to Bar Rivalry" *Oakland Tribune*, 15 September 1925.

94. Harry Hammond, "Delta Land Menaced by Mokelumne Water Claims of East Bay," *Stockton Independent*, 11 September 1925, p. 16.

95. "Tri-County Action to Save Water Flow for Mother Lode," *Stockton Record*, 29 September 1924, p. 10; "Asks Laws to Save Waters of the Valley," *Lodi Sentinel*, 20 November 1924, p. 1.

96. "Counties Plan to Reserve for Own Use Percentage of Water," *Stockton Record*, 17 October 1924, p. 17.

97. "Water Group Will Speed Up Election," *Lodi Sentinel*, 14 November 1925; "Water Users Seek County Organization," *Lodi Sentinel*, 11 November 1924; "Landowners Shown Need of Forming Water District," *Lodi Sentinel*, 9 January 1926. See also *Lodi Sentinel* editorials and articles: "Organize for Protection," 22 November 1924, p. 2; "Local Farmers Should Organize to Safeguard their Irrigation Rights," 27 September 1924, p. 14; "Save Waters of Mokelumne Urges Welch," 2 December 1924, p. 1.

98. "150 Protest Water Plan for District," *Lodi Sentinel*, 8 December 1925; "Mokelumne Water District Is Defeated," *Lodi Sentinel*, 6 May 1926; "Mason Sees Danger from Water Plans," *Lodi Sentinel*, 24 December 1925.

99. "150 Protest Water Plan for District," *Lodi Sentinel*, 8 December 1925, p. 1.

100. "Water Sale Authorized by District," *Oakland Tribune*, 6 June 1925, in *EBMUD 7/18–1/13/26* file, Oakland History Room, Oakland Public Library.

101. "Second Action Filed against East Bay Plan," *Lodi Sentinel,* 11 May 1926; *California Delta Farms, Inc. v. EBMUD,* 202 Cal. 793; *"Devine, et al. v. East Bay Municipal Utility District et al."* memorandum, 2 December 1932, *Thomas H. Means Papers,* 59, Water Resources Center Archive.

102. Other lawsuits did introduce important principles into California water law. In 1886, the judgment in *Lux v. Haggin,* 69 Cal. 255 (1886), finally resolved the conflicts between riparian and appropriated water rights by applying riparian rights only to those water claims made before the riparian lands came into private hands. This decision also confirmed appropriated rights on public lands. In 1926, the decision in *Herminghaus v. Southern California Edison Company,* 200 Cal. 81 (1926), permitted downstream landowners to block upstream projects that eliminated floods. This decision was instrumental in passing the 1928 legislation that established the "reasonable use" standard to eliminate waste in all water projects. Edward J. Horton, "The Organization of the Department of Water Resources of the State of California," M.A. thesis, Sacramento State College (1963), p. 9; Pisani, *From the Family Farm to Agribusiness,* p. 229.

103. *City of Lodi v. EBMUD,* 7 Cal. (2d) 316.

104. East Bay Municipal Utility District, *The Story of Water,* p. 7.

105. "Sewage A-M," 2, loose-leaf binder, East Bay Municipal Utility District Records Office, Oakland; Charles Gilman Hyde, Farnsworth Gray, and A. M. Rawn, *Report upon the Collection, Treatment and Disposal of Sewage and Industrial Wastes of the East Bay Cities, California* (n.p., 1941), p. 57, Crown Law Library, Stanford University.

106. Committee of East Bay Engineers, *Preliminary Report upon Sewage Disposal for the East Bay Cities of Alameda, Albany, Berkeley, El Cerrito, Emeryville, Oakland, Piedmont, San Leandro, and Richmond to the East Bay Executive Association* (Oakland: 1938), p. 7, Bancroft Library, Berkeley.

107. R. C. Kennedy, "Report on Sewage Collection and Disposal for Special District 1," p. 5, East Bay Municipal Utility District Records Office.

108. Hyde, Gray, and Rawn, *Report,* p. 57; B. I. Burnson to H. A. Knudsen, "Description of a Boat Trip Made to the Outfalls of the 77th Avenue, 85th Avenue and San Leandro Creek Sewers on May 21, 1945," 24 May 1945, in "Sewage A-M," East Bay Municipal Utility District Records Office.

109. Committee of East Bay Engineers, *Preliminary Report,* pp. 8–9.

110. Hyde, Gray, and Rawn, *Report,* 8.

111. Western Water Front Industries Association, "Temescal Creek and Chabot Road," *San Francisco Chronicle,* 28 October 1924; "West Oakland Sewer Branded Health Menace," *Oakland Tribune,* 19 June 1924, "Another Odor Problem," *Oakland Post-Enquirer,* 27 January 1937, all in *Oakland Sewers 1919–1930* file, Oakland History Room, Oakland Public Library.

112. California, State Board of Health, Bureau of Sanitary Engineering, "Appendix 1917," in *Report 1914–1916* (Sacramento, 1917), pp. 92, 102.

113. "East Bay Drainage Planned," *San Francisco Chronicle,* 14 February 1927, and "Sewer Lead Shift Needed," *Oakland Tribune,* 23 February 1927, both in *Oakland Sewers 1919–1930* file, Oakland History Room, Oakland Public Library.

114. "City Studies Cure for Odors on Bay Front," *Oakland Tribune,* 2 December 1936; "Sewer Merger to be Proposed," *Oakland Tribune,* 22 December 1936; "Oakland and

Emeryville in Sewage Plan," *Oakland Post-Enquirer,* 4 January 1937; "$10,000 Job Aid Fund Will Go to Sewer Projects," *San Francisco Examiner,* 22 May 1932; "Storm Sewer Work to Begin," *San Francisco Chronicle,* 21 February 1936, all in *Sewers 12/9/30–7/18/41* file, Oakland History Room, Oakland Public Library.

115. "Sewer Project Aid is Doubtful," *San Francisco News,* 8 July 1938; "Nine Cities Okeh Bid for PWA Finances," *San Francisco News,* 12 May 1938, both in *C. E. Grunsky Scrapbook,* "Clippings: Annexation, Litigation, Sewerage 1938," East Bay Municipal Utility District Records Office.

116. Committee of East Bay Engineers, *Preliminary Report,* p. 46; "There Will Be No Duplication of Sewer Costs," *San Leandro News,* 5 August 1938, in *C. E. Grunsky Scrapbook* "Clippings: Annexation, Litigation, Sewerage 1938"; "Utility District Accepts Plan for Garbage, Sewage Disposal," *Oakland Tribune,* 11 May 1944, in *East Bay Municipal Sewage District* file, Oakland History Room, Oakland Public Library; John S. Longwell, "Sewage Disposal for Special District No. 1 of the East Bay Municipal Utility District," delivered at the California Sewage Works Association Annual Convention, Monterey, California, 10 June 1946, p. 2, East Bay Municipal Utility District Records Office.

117. "Sewer Disposal Problem," *Oakland Tribune,* 12 February 1945, in "Sewage N–Z" loose-leaf binder, East Bay Municipal Utility District Records Office. During the war years, the East Bay population grew by 50 percent. Water use—and, therefore, sewage disposal—more than doubled from 1941 to 1945.

118. "Voters May Get Sewage District Plan," *Oakland Tribune,* 13 July 1944, and "Sewage Plan Protested," *Oakland Post-Enquirer,* 20 June 1944, both in *East Bay Sewage District* file, Oakland History Room, Oakland Public Library. "Organized Labor to Oppose Sewage Disposal Plan," *East Bay Labor Journal,* 20 October 1944, and "Take This to the Polls November 7th," CIO endorsement slip, November 1944(?), both in "Sewage N–Z," East Bay Municipal Utility District Records Office.

119. California, Department of Public Health, Bureau of Sanitary Engineering, "Report on Status of Disposal of Garbage and Sewage ... from the Standpoint of Stream Pollution," 1 January 1939, p. 11, Bancroft Library, Berkeley.

120. Gabriel Kolko, in *The Triumph of Conservatism: A Reinterpretation of American History, 1900–1916* (Chicago: Quadrangle Books, 1967), describes the Progressive Movement's goals in these economic terms. Samuel P. Hays' *Conservation and the Gospel of Efficiency: The Progressive Conservation Movement, 1890–1920* (Cambridge, Mass.: Harvard University Press, 1959) applies similar interpretations of Progressivism specifically to natural resource issues.

121. There is an extensive literature on the development and implications of California's two largest water projects. See, for example, Pisani, *From the Family Farm to Agribusiness.*

122. *Fitzgerald v. Orton,* 5 Cal. 308, 309 (1855) as quoted by Donald J. Pisani, "Enterprise and Equity: A Critique of Western Water Law in the Nineteenth Century," *Western Historical Quarterly* (January 1987), p. 21.

123. Walton Bean and James J. Rawls, *California: An Interpretive History* (New York: McGraw-Hill, 1983), p. 199; Arthur McEvoy, *The Fishermen's Problem: Ecology and Law in the California Fisheries, 1850–1980* (Cambridge: Cambridge University Press, 1986), p. 67; Richard J. Orsi, "The Octopus Reconsidered: The Southern Pacific and Agri-

cultural Modernization in California, 1865–1915," *California Historical Quarterly* 54 (1975), pp. 200–201. For a description of the flooding and river-bed silting caused by hydraulic mining, see Robert L. Kelley's *Battling the Inland Sea: American Political Culture, Public Policy, and the Sacramento Valley, 1850–1986* (Berkeley: University of California Press, 1989).

124. California Department of Public Works, Division of Engineering and Irrigation, "Bulletin 4—Water Resources of California: Report to the Legislature" (Sacramento: 1923), pp. 18, 39.

125. Gerald D. Nash, "Stages in California's Economic Growth, 1870–1970: An Interpretation," *California Historical Quarterly* 51 (1972), p. 319; Bureau of the Census. *Manufactures 1919*, vol. 9 of *Fourteenth Census of the United States taken in the Year 1920* (Washington, D.C.: 1923), p. 83; Bureau of the Census, "Bulletin: Agriculture: California," in *Fourteenth Census*, pp. 25–30.

126. William Kahrl estimates that late-twentieth-century industries use as much as 20 percent of the water delivered to California cities. In the early decades of the century, East Bay industries would have represented a higher percentage of local consumption, for several reasons. The East Bay housed a high concentration of water-intensive food-processing and oil-refining plants. Furthermore, East Bay residents in the 1920s did not have access to the washing machines, dishwashers, garbage disposals, lawn sprinklers, and swimming pools that have boosted late-twentieth-century domestic per capita water use to 200 gallons per day. Moreover, East Bay industry expanded rapidly between 1909 and 1930. From 1909 to 1920, the number of firms in the East Bay cities of Oakland, Berkeley, and Alameda increased by one-third (from 576 to 758 firms); the number of employees of these firms grew by nearly three-and-a-half-times (from 8,904 to 32,453 employees). By 1929, industries in Alameda and Contra Costa counties had increased in number by another two-thirds (to 1,229 firms) and employed a third more people (43,740 employees). William L. Kahrl, ed., *The California Water Atlas* (Sacramento: 1979), pp. 86–88; Nelson Blake, "Water and the City: Lessons from History," in Howard Rosen and Ann Keating, eds., *Water and the City: The Next Century* (Chicago: Public Works Historical Society, 1991), p. 66; Bureau of the Census, *Manufactures: 1919*, 83, 108–11; Bureau of the Census, *Manufactures: 1929*, vol. 3 in *Fifteenth Census of the United States* (Washington, D.C.: 1933), pp. 61–62.

127. John Walton, *Western Times and Water Wars: State, Culture and Rebellion in California* (Berkeley: University of California Press, 1992), pp. 169–75.

128. "Water Committee Plans to Open Fight on Mokelumne Demands of East Bay Dist.," *Lodi Sentinel*, 5 September 1925, p. 1.

129. Tom Sitton, "John Randolph Haynes and the Left Wing of California Progressivism," in William Deverell and Tom Sitton, eds., *California Progressivism Revisited* (Berkeley: University of California Press, 1994), p. 27.

130. "Governor to Ask Laws To Save Water," *Lodi Sentinel*, 28 August 1926, p. 1.

131. "Governor to Ask Laws To Save Water," *Lodi Sentinel*, 28 August 1926, p. 1.

132. Pisani, *From the Family Farm to Agribusiness*, pp. 394–98.

133. Samuel P. Hays discussed the Progressives' fascination with efficiency and conservation at length in *Conservation and the Gospel of Efficiency: The Progressive Conservation Movement, 1890–1920* (Cambridge: Harvard University Press, 1959).

134. "Conservation Planned For Water Flow," *Lodi Sentinel*, 13 January 1925, p. 1.
135. "Apportioning Mokelumne River Water," *Lodi Sentinel*, 14 February 1925, p. 2.
136. "The Fate of Municipal Ownership" *Pacific Rural Press*, 11 April 1925, p. 452; "Floating Islands of Government," *Pacific Rural Press*, 6 June 1925, p. 694.
137. "The Water and Power Act," *Stockton Record*, 28 October 1924, p. 18.
138. Leo A. McClatchy, "Reclamation Project Settlers Face Ejection Unless They Pay Up Millions Due to Government," *Stockton Record*, 11 September 1925, p. 1; "Failure of Public Ownership," *Lodi Sentinel*, 23 September 1924, p. 2; "The Fate of Municipal Ownership" *Pacific Rural Press*, 11 April 1925, p. 452; "Floating Islands of Government," *Pacific Rural Press*, 6 June 1925, p. 694; "The Melones Reservoir Plan," *Stockton Record*, 20 March 1925, p. 24.

Conclusion

1. See, for example, the Massachusetts State Board of Health report on north region sewerage, Massachusetts State Board of Health, *Report . . . upon the Sewerage of the Mystic and Charles River Valleys* (Boston: 1889), pp. 6–8, 20, and the East Bay studies of regional sewerage Committee of East Bay Engineers, *Preliminary Report*, pp. 10, 25–30. Financial considerations did not uniformly determine public service strategies. Cities could have filtered their sewerage and thus reduced the need to treat drinking water, but usually they left the problem of water purity to downstream communities. See Joel Tarr, "The Development and Impact of Urban Wastewater Technology: Changing Concepts of Water Quality Control, 1850–1930," in *Pollution and Reform in American Cities, 1870–1930*, edited by Martin V. Melosi (Austin: University of Texas Press, 1980), pp. 70–73, 77–78. For a discussion of the way infrastructure delayed the implementation of new policies, see Christine M. Rosen, *The Limits of Power: Great Fires and the Process of City Growth in America* (Cambridge: Cambridge University Press, 1986).
2. Frederick P. Stearns, engineer for the Wachusett system, had not only outlined the Swift and Ware Rivers projects in his original plans for Boston's regional waterworks, but also identified the Deerfield and Westfield Rivers as possible sources of supply. Massachusetts State Board of Health, *Report . . . upon a Metropolitan Water Supply* (Boston: 1895), p. xvi; Massachusetts General Court, *Legislative Hearings as to the Nashua River Water Supply before the Committees on Metropolitan Affairs and Water Supply*, 25 February–29 April 1895, p. 18, State House Library, Boston. For a detailed discussion of the history of the Quabbin reservoir, including the similarities between water supply decisions made in 1848, 1895, and 1926, see Fern L. Nesson, *Great Waters: A History of Boston's Water Supply* (Hanover, N.H.: University Press of New England, 1983) and Donald W. Howe, *Quabbin: the Lost Valley* (Ware, Mass.: 1951).
3. The Central Valley Project began as a state-sponsored irrigation network. The federal government adopted it during the Depression for financial reasons. In the 1960s, the state government began the State Water Plan to provide irrigation water without the acreage limits imposed federal reclamation programs. A number of authors have explored the development and implications of these projects in detail, including Donald Pisani, Norris Hundley, Marc Reisner, Robert Gottlieb, and Margaret FitzSimmons.

4. East Bay Municipal Utility District, "All About EBMUD" (Oakland, 1991), p. 2.

5. California State Lands Commission, *Delta-Estuary*, pp. 19–20, 34, 37, 45.

6. The Peripheral Canal referendum lost support in Southern California when amended to prohibit any further development of Northern California's rivers.

7. On opposition to Los Angeles' appropriations of water from the Mono Lake watershed, see John Walton, *Western Times and Water Wars: State, Culture and Rebellion in California* (Berkeley: University of California Press, 1992), pp. 264–68; and John Hart, *Storm over Mono: The Mono Lake Battle and the California Water Future* (Berkeley: University of California Press, 1996).

8. Glenn F. Bunting, "Bush OKs Water Policy Overhaul," *Los Angeles Times*, 31 October 1992.

9. John Cumbler, "Whatever Happened to Industrial Waste: Reform, Compromise and Science in Nineteenth-Century New England," unpublished manuscript, 1993, p. 29; "Sewage Disposal," *Bulletin of the League of American Municipalities* 2 (1904), p. 78.

10. Charles Haar, as quoted in Timothy G. Little, "Court Appointed Special Masters in Complex Environmental Litigation: City of Quincy v. Metropolitan District Commission," *Harvard Environmental Law Review* 8 (1984), pp. 448, 454.

11. Kathy Castagna, "Historical Perspectives and Overview," NOAA Estuary of the Month Seminar Series No. 4, *Boston Harbor and Massachusetts Bay: Issues, Resources, Status and Management*, 13 June 1985, Washington, D.C. (Washington, D.C.: U.S. Department of Commerce, 1985), pp. 5, 12.

12. Cheryl Breen, "Sewage Management," NOAA Estuary of the Month Seminar Series No. 4, *Boston Harbor and Massachusetts Bay: Issues, Resources, Status and Management*, 13 June 1985 (Washington, D.C.: U. S. Department of Commerce, 1985), p. 23.

13. Eric Jay Dolin, "Boston's Troubled Waters," *Environment* 34:6 (1992), p. 10. Many newspaper and magazine articles have traced the history of the Boston Harbor Cleanup. Two of the most thorough are Timothy G. Little, "Court Appointed Special Masters in Complex Environmental Litigation: City of Quincy v. Metropolitan District Commission," *Harvard Environmental Law Review* 8 (1984), pp. 435–75; and Kelly Slater, "The Port of Boston," *Sanctuary* (July/August 1986), pp. 3–7.

14. Little, "Court Appointed Special Masters," 443–44, 450–52, 466–67. The court held the hearing on this proposed injunction on 15 June 1983.

15. Donald R. Harleman, "Boston Harbor Cleanup: Use or Abuse of Regulating Authority?" (Boston: nd), pp. 8–9. A subsequent lawsuit by the federal Environmental Protection Agency has forced the Massachusetts Water Resources Authority to follow a court imposed clean-up schedule.

16. Massachusetts Water Resources Authority, "How We Can Clean the Dirtiest Harbor in America" (Boston: 1988?), p. 4; Massachusetts Water Resources Authority, *1987 Annual Report* (Boston: 1988), pp. 4–5.

17. East Bay Municipal Utility District, Office of Reclamation, "1992 Annual Report: Water Reclamation" (Oakland: 1992), p. 3; phone conversation with Patricia Mulvey, 12 August 1993. A $2 million feasibility study of this project is currently underway.

18. Within the cities, regional public works uniformly increased access to adequate water and sewerage regardless of race or class. This, indeed, is one of the greatest

triumphs of these projects. Outside metropolitan networks, disenfranchised groups have not fared so well; in the arid West, Native American and Hispanic communities have struggled for adequate water supplies. Although ethnic issues have played a smaller role in urban regionalism, some ethnic and class biases do appear in the willingness of California's cities to sacrifice commercial fisheries on the altar of inexpensive industrial waste disposal, as Arthur McEvoy documents in *The Fishermen's Problem: Ecology and Law in the California Fisheries, 1850–1980* (Cambridge: Cambridge University Press, 1986).

Bibliography

Newspapers

Berkeley Gazette
Berkeley Times
Boston Daily Globe
Boston Daily Journal
Boston Evening Transcript
Boston Globe
Boston Journal
Boston Morning Journal
Boston Pilot
Boston Post
Boston Sunday Herald
Bulletin of the League of American Municipalities
Clinton Daily Item
East Bay Labor Journal
Lodi News
Lodi Sentinel
Los Angeles Times
Oakland Bulletin
Oakland Daily Post
Oakland Enquirer
Oakland Post
Oakland Press
Oakland Tribune
Pacific Municipalities
Pacific Rural Press
Richmond Daily Independent
Richmond Independent
Richmond News
Richmond Record Herald
San Francisco Call
San Francisco Chronicle
San Francisco Enquirer
San Francisco News
San Leandro News
Stockton Independent
Stockton Record

Boston and Massachusetts Documents

Collections

Boston and Cambridge Water Supply. Pamphlet collection. Widener Library, Cambridge, Massachusetts.

Boston Water Supply, 1844–1845. Pamphlet collection. Widener Library, Cambridge, Massachusetts.

Boston Water Supply: Clippings from Boston Newspapers, 1895–1904. Rare Books, Boston Public Library, Boston, Massachusetts.

C. A. Cook Scrapbook. Beaman Memorial Library, West Boylston, Massachusetts.

A. E. Lindbury [Pillsbury], Republican Party Politics in Boston, 1873–82. Huntington Library, San Marino, California.

Municipalities: Water, Gas, Street Railways. Pamphlet collection. Widener Library, Cambridge, Massachusetts.

Papers Relating to Sewerage and Sewage Disposal. State Library, Boston.
Transcripts of Hearings before City Council 1910–1920. Boston Public Library, Boston.
Uncatalogued Boston City Records Collection. Boston Public Library, Boston.
John C. Warren Papers. Massachusetts Historical Society, Boston.
Water Reports 1835–1845. Pamphlet Collection. Massachusetts Water Resources Authority Library, Boston.

Other Sources

Baldwin, Laommi. *Report on the Subject of Introducing Pure Water into the City of Boston.* Boston: 1834.
Boston. *A Catalogue of the City Councils of Boston 1822–1908, Roxbury 1846–1867, Charlestown 1847–1873 . . . .* Boston: 1909.
______. *Digest of Laws and Ordinances Relating to the Public Health.* Boston: 1869.
______. *Digest of Laws and Ordinances Relating to the Public Health.* Boston: 1873.
______. *Documents of the City of Boston.* 1850–1895.
______. *Reports and Proceedings of the City Council of Boston.* 1860–1896.
Boston Board of Health. *A Communication . . . to the Committee on Improved Sewage.* Boston: 1876.
______. *Ordinances Prescribing Rules and Regulations Relative to Nuisances, Sources of Filth, and Causes of Sickness within the City of Boston.* Boston: 1833.
Boston City Council. Committee on Sewers. *Report.* Boston: 1874.
______. Joint Special Committee . . . to Consider . . . a Public Park. *Report and Accompanying Statements and Communications Relating to a Public Park for the City of Boston.* Boston: 1869.
______. *Report of the Joint Special Committee on Improved Sewerage.* Boston: 1877.
Boston Engineering Department. *Charles River Pollution.* Boston: 1892.
Boston Health Department. *Manuals of Revised Laws and City Ordinances Relating to Public Health.* Boston: 1904.
Boston Sewer Department. *Annual Reports of the Superintendent of Sewers.* 1860–1899.
Boston Water Board. *Boston Water Works: Additional Supply from the Sudbury River.* Boston: 1882.
______. *Special Report . . . to the City Council, September 6, 1883.* Boston: 1883.
Boston Water Commission. 4 January 1850. *Final Report.* Boston: 1850.
Boston Water Supply Department. *Annual Reports.* 1890–1895.
Chesbrough, E. S.; Lane, Moses; and Folsom, Charles Follen. *The Sewerage of Boston, a Report by a Commission.* Boston: 1876.
Choate, Charles F., Jr. "Brief and Argument in Behalf of the Clinton Business Men's Association before a Special Joint Committee to Investigate Violations of Law on Wachusett Reservoir." np. State Library, Boston.
Curtis, Josiah. *Brief Remarks on the Hygiene of Massachusetts, but More Particularly the Cities of Boston and Lowell, Being a Report to the American Medical Association.* Philadelphia: 1849.
Eddy, Robert H. *Report on the Introduction of Soft Water into the City of Boston.* Boston: 1836.

Emerson, George B. *Remonstrance of George B. Emerson and Other Taxpayers of Boston, against the Adoption of the System of Sewerage Proposed in Report No. 3.* Boston: 1876.

Folsom, Charles Follen. *"Metropolitan Main Drainage," Remarks before the Joint Committee on Improved Sewerage, City Hall, May 9, 1876.* Boston: 1876.

———. "The Present Aspect of the Sewage Question as Applied to Boston: A Paper Read before the American Statistical Association, Boston, April 20, 1877." Boston: 1877. State Library, Boston.

The Inagural Addresses of the Mayors of Boston. Boston: 1894.

Joint Special Committee of the [Massachusetts] House of Representatives. "Sanitary Survey of the State." Boston: 1849.

Legislative Packets for Massachusetts Acts and Resolves, Massachusetts State Archives, Boston.

Lyman, Theodore, Jr. *Communication to the City Council, on the Subject of the Introduction of Water into the City.* Boston: 1834.

Massachusetts. *Acts and Resolves Passed by the General Court of Massachusetts.* Boston: 1881.

———. *Journal of the House of Representatives.* Boston: 1870–1900.

———. *Journal of the Senate.* Boston: 1870–1900.

———. *Legislative Hearings as to the Nashua River Water Supply before the Committees on Metropolian Affairs and Water Supply.* 25 February–29 April 1895. State Library, Boston.

Massachusetts Bays Program. *Massachusetts Bays 1991 Comprehensive and Management Plan: An Evolving Plan for Action.* Boston: 1991.

Massachusetts Committee on Public Health. *Evidence and Arguments on Petitions of Cambridge and Boston for Leave to Take Water from Shawshine River.* Boston: 1882. Widener Library, Cambridge.

Massachusetts General Court. *Documents Printed by Order of the Senate of the Commonwealth of Massachusetts.* Boston: 1845–1900.

———. *Documents of the House of Representatives.* Boston: 1845–1900.

———. *Hearing on a Bill Relative to the Protection of the Public Health in the Valley of the Neponset River,* 9 May 1900. State Library, Boston.

Massachusetts Legislative Research Council. *Report Relative to Water Shortage and Industrial Water Use and Reuse.* Boston: 1966.

Massachusetts Medical Commission. *The Sanitary Condition of Boston: The Report of a Medical Commission to the Boston Board of Health.* Boston: 1875.

Massachusetts Metropolitan Area Planning Council. Camp Dresser & McKee, consulting engineers. *Alternative Regional Sewerage Systems for the Boston Metropolitan Area.* Boston: 1972.

Massachusetts Metropolitan Drainage Commissioners. *Report on the Mystic and Charles River Valleys Drainage.* Boston: 1882.

Massachusetts Metropolitan Sewerage Commissioners. *Main Drainage Works of Boston and Its Metropolitan Sewerage District.* Boston: 1899. Widener Library, Harvard University.

———. *Report ... upon a High-Level Gravity Sewer for the Relief of the Charles and Neponset River Valleys.* Boston: 1899.

Massachusetts State Board of Health. *Annual Reports.* 1870–1878, 1887–1895.

______. *Report . . . upon the Sewerage of the Mystic and Charles River Valleys.* Boston: 1889. Boston Public Library, Boston.

______. *Report of the State Board of Health upon the Discharge of Sewage into Boston Harbor.* Boston: 1900.

Massachusetts State Board of Health, Lunacy and Charity. *Annual Reports.* 1879–1886.

______. *Report . . . upon the Discharge of Sewage into Boston Harbor.* Boston: 1900.

Massachusetts Water Resources Authority. *1987 Annual Report.* Boston: 1988.

______. Environmental Quality Department. Menzie-Cura & Associates, Inc. *Boston Harbor: Estimates of Loadings.* Technical Report 91–4. Boston: 1991.

______. "How Can We Clean the Dirtiest Harbor in America?" Boston: MWRA, 1988.

Matthews, Nathan, Jr. *The City Government of Boston.* Boston: 1895.

Medford. *Report of the Board of Selectmen on the Subject of Sewerage.* Medford, Mass.: Chronicle Press, 1873.

Nichols, William Ripley, and Derby, George. *Sewerage; Sewage; the Pollution of Streams; the Water-Supply of Towns: A Report to the State Board of Health of Massachusetts.* Boston: 1873.

Quincy Chamber of Commerce. "An Appeal to the Members of the General Court of Massachusetts: Give Your Citizens the Joy of Clean Waters." Quincy, Mass.: 1939.

Report of a Commission Appointed to Consider a General System of Drainage for the Valleys of the Mystic, Blackstone and Charles Rivers. Boston: 1886.

Roxbury Commission on Annexation. *Report of the Commissioners . . . of the City of Roxbury and Boston, respectively, on the Union of the Two Cities . . .* Roxbury, Mass.: 1867.

Stearns, Frederick P. "Information Relating to the Claims of the Town of Clinton," 23 February 1901. State Library, Boston.

Treadwell, Daniel. "Report Made to the Mayor and Aldermen of the City of Boston, on the Subject of Supplying the Inhabitants of that City with Water." Boston: 1925.

Waring, George E. *Irvington Sanitary Survey, and Examination as to Local Causes of Fire and Ague . . .* New York: 1879. State Library, Boston.

______. *Storm Water in Town Sewerage*. Newport, R.I.: 1881. Boston Public Library.

West Boylston's Argument in Regard to the Metropolian Water Supply. Worcester, Mass.: O. B. Wood, 1895.

Winslow, C. E. A., and Phelps, Earl B. *Investigations on the Purification of Boston Sewage.* Water Supply and Irrigation Papers, U.S. Geological Survey, Department of Interior, Series L. Washington, D.C.: Government Printing Office, 1906.

East Bay and California Documents

Collections

David Prescott Barrows Papers. Bancroft Library, Berkeley, California.

Louis Bartlett Papers. Bancroft Library, Berkeley, California.

Berkeley, California: Letters from Citizens (1943). Bancroft Library, Berkeley, California.

Clippings Files on East Bay Municipal Utility District and Oakland. Oakland History Room, Oakland Public Library, Oakland, California.
Commonwealth Club of California, Records of the City Planning Section. Bancroft Library, Berkeley, California.
East Bay Municipal Utility District History Files and Scrapbooks. East Bay Municipal Utility District Records Office, Oakland, California.
Bernard A. Etcheverry Papers. Water Resources Center Archive, Berkeley, California.
John D. Galloway Papers. Water Resources Center Archive, Berkeley, California.
C. E. Grunsky Scrapbooks. East Bay Municipal Utility District Records Office, Oakland, California.
Frank C. Havens Papers. Bancroft Library, Berkeley, California.
Charles Gilman Hyde Papers. Water Resources Center Archive, Berkeley, California.
John Edmund McElroy Collection. Bancroft Library, Berkeley, California.
Sidney T. Harding Papers. Water Resources Center Archive, Berkeley, California.
Thomas H. Means Papers. Water Resources Center Archive, Berkeley, California.
Oakland Collections. Bancroft Library, Berkeley, California.
George Pardee Scrapbooks. Special Collections, Stanford University Library, Stanford, California.

Other Sources

Adams, Arthur J. "How Shall Oakland Secure Public Ownership of Water Works?" np, 1904. Water Resources Center Archive, Berkeley.
Alameda City Manager. *Annual Reports.* 1915–1930.
"Bay Cities Water Problems." *Transactions of the Commonwealth Club* 12 (1915).
Bowhill, Thomas. "The Alvarado Artesian Water of the Oakland Water Company Compared to the Surface Waters of Lake Temescal and Lake Chabot of the Countra Costa Water Company." Oakland: 1895.
California. *Constitution of California as of 1879 as Amended and in Effect February 20, 1934.* Sacramento, 1934.
California Commission on County Home Rule. *Final Report.* Sacramento: 1931.
California Department of Public Health. "Report on Status of Disposal of Garbage and Sewage . . . from the Standpoint of Stream Pollution" (1 January 1939). Sacramento: 1939.
California Department of Public Works. "Bulletin 4—Water Resources of California: Report to the Legislature." Sacramento: 1923.
______. *Feasibility of the Reclamation and Conveyance of Sewage from San Francisco, Peninsula Cities and Communities and San Jose, for Use in Santa Clara County.* Sacramento: 1951.
California Joint Federal-State Water Resources Comission. *Report.* np, 1930.
California Legislature. *Final Calendars of Legislative Business.* Sacramento: 1910–1930.
______. Assembly. *Journal of the Assembly.* Sacramento: 1910–1930.
______. Assembly Metropolitan Advisory Council. *Transcript of Proceedings: Multipurpose Metropolitan Districts.* Sacramento: 1961.

______. Senate. *Journal of the Senate.* Sacramento: 1910–1930.

California State Lands Commission. *Delta-Estuary: California's Inland Coast: A Public Trust Report.* Sacramento: 1991.

California State Reconstruction and Reemployment Commission. *Special Reports, 1944–1945.* Sacramento: 1945.

California State Water Problems Conference. *Report* (25 November 1916). Sacramento: 1916.

Chamberlain, R. H.; Howard, John L.; Olney, Warren; Kahn, Sol; and Taylor, James P. *Municipal Ownership of Water and Available Sources of Supply for Oakland, California.* Oakland: Enquirer, 1903.

Committee of East Bay City Engineers. *Preliminary Report upon Sewage Disposal for the East Bay Cities of Alameda, Albany, Berkeley, El Cerrito, Emeryville, Oakland, Piedmont, San Leandro, and Richmond to the East Bay Executives Association.* Oakland: 1938.

Davie, John L. *His Honor, the Buckaroo: The Autobiography of John L. Davie,* ed. by Jack W. Herzberg. Reno, Nev.: Jack Herzberg, 1988.

______. *Mayor's Message.* Oakland: 1929.

Davis, Arthur Powell; Goethals, George W; and Mulholland, William. *Additional Water Supply of East Bay Municipal Utility District: A Report to the Board of Directors.* Oakland: 1924.

Dockweiler, J. H. *General Information Regarding Proposed Metropolitan Municipal Water District, Alameda County, California.* Bancroft Library, Berkeley.

______. *Report on Sources of Water Supply: East Region of San Francisco Bay.* San Francisco: 1912.

East Bay Municipal Utility District. "Application for Loan for Extension of Water Supply Purification and Distribution Systems." np, 1931. Water Resources Center Archive, Berkeley.

______. *Chronology and Principal Features of the Mokelumne Water Supply Project and Distribution System.* Oakland(?): n.d. Bancroft Library, Berkeley.

______. *East Bay Sewers: Past, Present, and Future.* Oakland: 1983.

______. "History and Structure of the East Bay Municipal Utility District." 1969. East Bay Municipal Utility District Records Office, Oakland.

______. "History of the Water Supply for the East Bay Communities." 1920. East Bay Municipal Utility District Records Office, Oakland.

______. "The History of Water and Water Companies in the East Bay Area." n.d. East Bay Municipal Utility District Records Office, Oakland.

______. Special District No. 1. [Kennedy, R. C.]. *Report: Engineering and Construction of the East Bay Sewage Disposal Project for Albany, Berkeley, Emeryville, Oakland, Piedmont and Alameda, California. June 1954.* Oakland: 1954.

______. *The Story of Water: A Brief History of the East Bay Municipal Utility District and a Description of the Source, Transmission, Treatment and Distribution of the Water Supply to the Cities within Its Boundaries.* Oakland: 1931. Bancroft Library, Berkeley.

Gibson, Mary S., comp. *A Record of Twenty-five Years of the California Federation of Women's Clubs, 1900–1928.* California Federation of Women's Clubs, 1927.

Gillespie, Chester C. "Origins and Early Years of the Bureau of Sanitary Engineering," interviewed by Malca Chall. Berkeley: Regional Oral History Office, 1971. Bancroft Library, Berkeley.

Goodall, Merrill; Sullivan, John D.; and De Young, Tim. *Water Districts in California: An Analysis by Type of Enabling and Political Decision Process.* Unpublished study for the California Department of Water Resources, March 1977.

Grunsky, C. E., and Manson, Marsden. "Report on the Water Supply of San Francisco, California, 1900 to 1908, Inclusive." San Francisco: 1909.

Harroun, Philip E. *Report to the Commission of the East Bay Cities on the Water Supply for the Cities of Oakland, Berkeley, Alameda and Richmond.* np, 1920. Bancroft Library, Berkeley.

Hyde, Charles Gilman; Gray, Harold Farnsworth; and Rawn, A. M. *Report upon the Collection, Treatment and Disposal of Sewage and Industrial Wastes of the East Bay Cities, California.* 1941. Crown Law Library, Stanford University, Palo Alto, California.

Jenks & Adamson, and Kennedy Engineers. *East Bay Dischargers Water Quality Management Program: Final Report.* np, 1972. Water Resources Center Archive, Berkeley.

"Joint Hearing before the Division of Water Rights, Department of Public Works, State of California and the Federal Power Commission," 11 September 1925. Recorder's transcript. East Bay Municipal Utility District Records Office, Oakland.

Kennedy, R. C. *Report on Sewage Collection and Disposal for Special District 1.* Oakland: 1946. East Bay Municipal Utility District Records Office, Oakland.

Longwell, John S. "Sewage Disposal for Special District No. 1 of the East Bay Municipal Utility District." Presented at the California Sewage Works Association annual meeting, Monterey, California, 10 June 1946. East Bay Municipal Utility District Records Office, Oakland.

Manson, Marsden, and Grunsky, C. E. *Report on the Plans and Estimates for the Proposed Main Sewers East of Lake Merritt.* Oakland: 1893. Bancroft Library, Berkeley.

Oakland Board of Engineers. *Report . . . on the Grades, Streets and Sewerage of the City of Oakland.* Oakland: 1870.

Oakland Chamber of Commerce. *Report on the East Bay Water Problem.* Oakland: 1923. East Bay Municipal Utility District Records Office, Oakland.

Oakland City Council. *Minutes.* 1915–1925.

______. "Resolution of Intention: Municipal Improvements." Oakland: 1902.

Oakland Mayor. "Annual Message." Anson Barstow. Oakland: 1902.

Oakland Park Commission. *The Park System of Oakland, California.* Oakland: 1910.

Sacramento Citizens' Water Investigation Committee. *Report.* Sacramento: 1901. Stanford University Library, Palo Alto, California.

San Francisco Bureau of Engineering. *Hetch Hetchy Water Supply.* San Francisco: 1925.

Additional Primary Materials

Chadwick, Edwin. *Report on the Sanitary Condition of the Labouring Population of Great Britain.* 1842.

Chesbrough, E. S. *Report of the Results of Examinations Made in Relation to Sewerage in Several European Cities in the Winter of 1856–1857.* Chicago: 1858.

Goodnough, X. H. *Preliminary Reports on the Disposal of New York's Sewage.* New York: 1914.

NOAA Estuary of the Month Seminar Series. *Boston Harbor and Massachusetts Bay: Issues, Resources, Status and Management, Washington, D.C., 13 June 1985.* Washington, D.C.: U.S. Department of Commerce, 1985.

United States Environmental Protection Agency. *Draft Environmental Impact Statement on the Upgrading of the Boston Metropolitan Area Sewerage System.* Washington, D.C.: 1978.

United States Senate. *Hearings re: Senate Bill 2610: Proposed Right of Way for Hetch Hetchy Reservoir.* np, 1913. East Bay Municipal Utility District Archives, Oakland.

Secondary Materials

Anderson, Letty. "Hard Choices: Supplying Water to New England Towns." *Journal of Interdisciplinary History* 15 (1984): 211–34.

Angel, Arthur Desko. "Political and Administrative Aspects of the Central Valley Project of California." Ph.D. dissertation, University of California Los Angeles, 1944.

Ashworth, William. *Nor Any Drop to Drink.* New York: Summit Press, 1982.

Atkins, Gordon. "Health, Housing and Poverty in New York City, 1865–1898." Ph.D. dissertation, Columbia University, 1947.

Bagwell, Beth. *Oakland: The Story of the City.* Novato, Calif.: Presidio Press, 1982.

Bailie, William. *An Inquiry into Boston's Water Supply and its Relation to the Public Health, with Some Startling Facts about the Pollution of Sources of Supply.* Boston: 1909.

Barth, Gunther. *Instant Cities: Urbanization and the Rise of San Francisco and Denver.* New York: Oxford University Press, 1975.

Bartlett, Louis. *Memoirs.* Berkeley: University of California Regional Cultural History Program, 1957.

Bean, Walton, and Rawls, James J. *California: An Interpretive History.* New York: McGraw-Hill, 1983.

Bernard, Richard M., and Rice, Bradley R. "Political Environment and the Adoption of Progressive Municipal Reform." *Journal of Urban History* 1 (1975): 149–74.

Blackford, Mansel G. *Politics of Business in California, 1890–1920.* Columbus, Ohio: Ohio State University Press, 1977.

Blake, John B. *Public Health in the Town of Boston, 1630–1822.* Cambridge, Mass.: Harvard University Press, 1959.

Blake, Nelson M. *Water for the Cities: A History of the Urban Water Supply Problem in the United States.* Syracuse, N.Y.: Syracuse University Press, 1956.

Bollens, John C. *Exploring the Metropolitan Community.* Berkeley: University of California Press, 1964.

———. *Special District Governments in the United States.* Berkeley: University of California Press, 1957.

Borcherding, Thomas, ed. *Budgets and Bureaucrats: The Sources of Government Growth.* Durham, N.C.: Duke University Press, 1977.

Bowers, Marth H., and Carolan, Jane. *The Water Supply System of Metropolitan Boston, 1845–1947.* Wellesley, Mass.: Louis Berger, 1985.

Boyer, M. Christine. *Dreaming the Rational City: The Myth of American City Planning.* Cambridge, Mass.: MIT Press, 1983.

Boyer, Paul S. *Urban Masses and Moral Order in America, 1820–1920.* Cambridge, Mass.: Harvard University Press, 1978.

Bradlee, Nathanial J. *History of the Introduction of Pure Water into the City of Boston.* Boston: 1868.

Brock, William. *Investigation and Responsibility: Public Responsibility in the United States, 1865–1900.* Cambridge: Cambridge University Press, 1984.

Brown, M. Craig, and Halaby, Charles N. "Machine Politics in America, 1870–1945." *Journal of Interdisciplinary History* 17 (1987): 587–612.

Brownell , Blaine A., and Goldfield, David R. *The City in Southern History: The Growth of Urban Civilization in the South.* South Port Washington, N.Y.: Kennikat Press, 1977.

Burgess, Sherwood D. "Early History of the Oakland Water Supply, 1850–1876." Ph.D. dissertation, University of California, 1940.

______. "Oakland's Water War." *California History* (1985): 34–41.

______. *Water King: Anthony Chabot, His Life and Time.* Davis, Calif.: Panorama West Publishing, 1992.

Burns, Nancy. *Formation of American Local Governments: Private Values in Public Institutions.* New York: Oxford University Press, 1994.

Butler, Lynda L. "Environmental Water Rights: An Evolving Concept of Public Property." *Virginia Environmental Law Review* 9(2) (1990): 323–80.

Cain, Louis P. *Sanitation Strategy for a Lakefront Metropolis: The Case of Chicago.* DeKalb, Ill.: Northern Illinois University Press, 1978.

______. *The Search for an Optimum Sanitation Jurisdiction: The Metropolitan Sanitary District of Greater Chicago, A Case Study.* Essays in Public Works History, no. 10 (1980).

______. "Raising and Watering a City: Ellis Sylvester Chesbrough and Chicago's First Sanitation System." In Judith Walzer Leavitt and Ronald L. Numbers, eds., *Sickness and Health in America: Readings in the History of Medicine and Public Health* (Madison: University of Wisconsin Press, 1985), pp. 439–50.

Cazzaza, Joseph. "Boston Still Faces a Pollution Control Problem." *Journal of Water and Wastes Engineering* 7 (1970): 44–47.

Chicago Department of Public Works. *Chicago Public Works: A History.* Chicago: 1970.

Chudacoff, Howard. *The Evolution of American Urban Society.* Englewood Cliffs, N.J.: Prentice-Hall, 1981.

Clark, Eliot C. *Main Drainage Works of the City of Boston.* Boston: 1885.

Clark, Walter. "Public Ownership the Only Solution." *Pacific Municipalities* 37 (1923) 3–8.

Cleary, Edward J. *The Orsanco Story: Water Quality Management in the Ohio Valley under an Interstate Compact.* Baltimore: Johns Hopkins University Press, 1967.

Clinton Centennial Volume, 1850–1950. Clinton, Mass.: 1951.

Conkey, John Houghton, and Conkey, Dorothy Dunham. *The History of Ware, Massachusetts, 1911– 1960.* Ware, Mass.: 1961.

Conuel, Thomas. *Quabbin: The Accidental Wilderness.* Brattleboro, Vt.: Stephen Greene Press, 1981.

Cooper, Erwin. *Aqueduct Empire: A Guide to Water in California, Its Turbulent History and Its Management Today.* Glendale, Calif.: Arthur H. Clarke, 1968.

Corbridge, James N., Jr., ed. *Special Water Districts: Challenge for the Future.* Proceedings of the Workshop on Special Water Districts . . . np, 1983.

Cory, H. T. "Water Supply of the San Francisco–Oakland Metropolitan District with Discussion by F. T. Robson, et al." *Transactions* 50(1916): 1–43.

Cranz, Galen. *The Politics of Park Design: The History of Urban Parks in America.* Cambridge, Mass.: MIT Press, 1982.

Cromwell, Adelaide M. *The Other Brahmins: Boston's Black Upper Class, 1750–1950.* Fayetteville: University of Arkansas Press, 1994.

Cronon, William. *Nature's Metropolis: Chicago and the Great West.* New York: W. W. Norton, 1992.

Crouch, Winston W., and Dinerman, Beatrice. *Southern California Metropolis: A Study in Development of Government for a Metropolitan Area.* Berkeley: University of California Press, 1963.

Crouch, Winston W., and McHenry, Dean E. *California Government: Politics and Administration.* Berkeley: University of California Press, 1954.

Crouch, Winston W.; McHenry, Dean E.; Bollens, John C.; and Scott, Stanley. *California Government and Politics.* Englewood Cliffs, N.J.: Prentice-Hall, 1960.

Cumbler, John. "The Early Making of an Environmental Consciousness: Fish, Fisheries Commissions, and the Connecticut River." *Environmental History Review* 15 (1991): 73–92.

______. "Whatever Happened to Industrial Waste: Reform, Compromise and Science in Nineteenth-Century New England." American Society for Environmental History meeting, Pittsburgh, March 1993.

Daland, Robert T. *Government and Health: The Alabama Experience.* Birmingham: University of Alabama, 1955.

Davis, Mike. *City of Quartz: Excavating the Future in Los Angeles.* New York: Verso Press, 1990.

Delgado, James P. "'The Humblest Cottage Can in Short Time Afford . . . Pure and Sparkling Water': Early Efforts to Solve Gold Rush San Francisco's Water Shortage." *Pacific Historian* 26 (1982): 26–39.

Delmatier, Royce D.; McIntosh, Clarence F.; and Waters, Earl G. *The Rumble of California Politics, 1848–1970.* New York: John Wiley & Sons, 1970.

Deverell, William. "Building an Octopus: Railroad and Society in Late Nineteenth Century." Ph. D. dissertation, Princeton University, 1989.

______. *Railroad Crossing: Californians and the Railroad, 1850–1910.* Berkeley: University of California Press, 1994.

Deverell, William, and Sitton, Tom. *California Progressivism Revisited.* Berkeley: University of California Press, 1994.

Dolin, Eric Jay. "Boston's Troubled Waters," *Environment* 34:6 (1992), pp. 7–33.

______. *Dirty Water/Clean Water.* MIT Sea Grant College Program, Boston: 1990.

Donahue, Brian. "'Damned at Both Ends and Cursed in the Middle': The 'Flowage' of the Concord River Meadows, 1798–1862." *Environmental Review* 13 (1989): 47–68.

Donahue, Martin. "History of Administrative Law in Massachusetts, 1629–1932." *Journal of Legal History* 8 (1987): 330–66.

Dubos, René, and Dubos, Jean. *The White Plague: Tuberculosis, Man and Society*. Boston: Little, Brown, 1952.

Duffy, John. *A History of Public Health in New York City*. New York: Russell Sage Foundation, 1968.

East, Dennis. "Health and Wealth: Goals of the New Orleans Public Health Movement, 1879–84." *Louisiana History* 9 (1968): 245–75.

Einhorn, Robin L. *Property Rules: Political Economy in Chicago, 1833–1872*. University of Chicago Press, 1991.

Emanuels, George. *California's Contra Costa County: An Illustrated History*. Fresno, Calif.: Panorama West Books, 1986.

Emenhiser, JeDon A., ed. *Rocky Mountain Urban Politics*. Logan: Utah State University Press, 1971.

Engelbert, Ernest A., ed. *Metropolitan California: Papers Prepared for the Governor's Commission on Metropolitan Area Problems*. Sacramento: 1961.

Eshleman, John M. "The Regulation of Public Utilities." *Pacific Municipalities* 26 (1912), pp. 345–50.

Fitzgerald, Desmond. *History of the Boston Water Works from 1868 to 1876*. Boston: 1876.

———. *A Short Description of the Boston Waterworks*. Boston: 1895.

Fogelson, Robert W. *Fragmented Metropolis: Los Angeles 1850–1930*. Cambridge, Mass.: Harvard University Press, 1967.

Folk-Williams, John A.; Fry, Susan C.; and Hilgendorf, Lucy. *Western Water Flows to the Cities. Water in the West,* vol. 3. Santa Fe, N.M.: Western Network, 1985.

Ford, Kristina Louise Hensley. "Regional Association and Dissociation in the San Francisco Bay Area." Ph.D. dissertation, University of Michigan, 1976.

Formisano, Ronald P., and Burns, Constance K., eds. *Boston, 1700–1980: The Evolution of Urban Politics*. Westport, Conn.: Greenwood Press, 1984.

Frederick, Kenneth D., and Gibbons, Diana C., eds. *Scarce Water and Institutional Change*. Washington, D.C.: Resources for the Future, 1986.

Galambos, Louis. "Technology, Political Economy, and Professionalization: Central Themes of the Organizational Synthesis." *Business History Review* 57 (1983): 471–92.

Galishoff, Stuart. "Drainage, Disease, Comfort and Class: A History of Newark's Sewers." *Societas: A Review of Social History* 6 (1976): 121–39.

———. *Newark: The Nation's Unhealthiest City, 1832–1895*. New Brunswick, N.J.: Rutgers University Press, 1988.

Gillette, Howard, Jr., and Miller, Zane L., eds. *American Urbanism: An Historiographic Review*. New York: Greenwood Press, 1987.

Gluck, Peter R., and Meister, Richard J. *Cities in Transition: Social Changes and Institutional Responses in Urban Development*. New York: New Viewpoints, 1979.

Goldman, Joanne Mara. "The Development of a Sewer System in New York City, 1800–1866: Evolution of a Technological and Managerial Infrastructure." Ph.D. dissertation, SUNY Stony Brook, 1988.

Gottlieb, Robert, and Fitzsimmons, Margaret. *Thirst for Growth: Water Agencies as Hidden Governments in California*. Tucson: University of Arizona Press, 1991.

Goubert, Jean Pierre. *The Conquest of Water: The Advent of Health in the Industrial Age*, trans. by Andrew Wilson. Princeton, N.J.: Princeton University Press, 1986.

Grassman, Curtis E. "Prologue to California Reform: The Democratic Impulse, 1886–1898." *Pacific Historical Review* 42 (1973): 518–36.

Greene, J. R. *The Creation of Quabbin Reservoir: The Death of the Swift River Valley*. Athol, Mass.: Transcript Press, 1981.

Greer, Scott A. *Governing the Metropolis.* New York: John Wiley & Sons, 1962.

Guillerme, Andre. "The Genesis of Water Supply, Distribution and Sewerage Systems in France, 1800–1850." In Joel Tarr and Gabriel Dupuy, eds. *Technology and the Rise of the Networked City* (Philadelphia: Temple University Press, 1988), pp. 91–115.

Hall, D. J. "State Regulation or Municipally Owned Utilities." *Pacific Municipalities* 37 (1923), pp. 223–24.

Hall, Peter. *Urban and Regional Planning.* New York: John Wiley & Sons, 1975.

Hamilton, Helen M. "The History of West Boylston." Ph.D. dissertation, Clark University, 1954.

Hamlin, Christopher. *A Science of Impurity: Water Analysis in Nineteenth Century Britain.* Berkeley: University of California Press, 1990.

Hammack, David C. *Power and Society: Greater New York at the Turn of the Century*. New York: Russell Sage Foundation, 1982.

Handlin, Oscar. *Boston's Immigrants: A Study in Acculturation*, rev. ed. Cambridge, Mass.: Belknap Press, 1979.

Handlin, Oscar, and Burchard, John, eds. *The Historian and the City*. Cambridge, Mass.: MIT Press and Harvard University Press, 1963.

Hanson, Warren D. *A History of the Municipal Water Department and Hetch Hetchy System.* City and County of San Francisco, San Francisco: 1985.

Harding, S. T. *Water in California.* Palo Alto, Calif.: N-P Publications, 1960.

Harleman, Donald R. "Boston Harbor Cleanup: Use or Abuse of Regulatory Authority?" Boston: nd. Massachusetts Water Resources Authority Library.

Hart, John. *Storm over Mono: The Mono Lake Battle and the California Water Future.* Berkeley: University of California Press, 1996.

Hartog, Hendrik. *Public Property and Private Power: The Corporation of the City of New York in American Law, 1730–1870.* Chapel Hill: University of North Carolina Press, 1983.

Hays, Samuel P. *Conservation and the Gospel of Efficiency: The Progressive Conservation Movement, 1890–1920.* Cambridge, Mass.: Harvard University Press, 1959.

Hichborn, Franklin. *California Politics, 1891–1939.* Los Angeles: Haynes Foundation (nd).

Higgs, Robert. *Crisis and the Leviathan: Critical Episodes in the Growth of American Government.* New York: Oxford University Press, 1987.

Hooper, John H. "Sewage in Mystic River." *Medford Historical Register* 23 (1920): 45–53.

Hope, Barney, and Sheehan, Michael. "Political Economy of Centralized Water Supply in California." *Social Science Journal* 20(1983): 29–39.

Hope, Holly. *Garden City: Dreams in a Kansas Town.* Norman: University of Oklahoma Press, 1988.

Horton, Edward Joseph. "The Organization of the Department of Water Resources of the State of California." M.A. thesis, Sacramento State College, 1963.

Howe, Donald W. *Quabbin: The Lost Valley.* Quabbin Book House,Ware, Mass.: 1951.

Hughes, Thomas P. *Networks of Power: Electification in Western Society, 1880–1930.* Baltimore: Johns Hopkins University Press, 1983.

Hundley, Norris, Jr. *The Great Thirst: Californians and Water, 1770s–1990s*. Berkeley: University of California Press, 1992.

Hunt, Rockwell D., ed. *California and Californians*. vol. 3. San Francisco: Lewis, 1926.

Hurley, Andrew. "Creating Ecological Wastelands: The Case of Oil Pollution in New York City." American Society for Environmental History meeting, Pittsburgh, March 1993.

———. "The Social Biases of Environmental Change in Gary, Indiana, 1945–1980." *Environmental Review* 12 (1988): 1–21.

Hutchinson, W. H. "Prologue to Reform: The California Anti-Railroad Republicans, 1899–1905." *Southern California Quarterly* 44 (1962): 175–218.

———. "Southern Pacific: Myth and Reality." *California Historical Society Quarterly* 48 (1969): 325–34.

Issel, William. "Liberalism and Urban Policy in San Francisco from the 1930s to the 1960s." *Western Historical Quarterly* 22 (1991): 431–50.

Jackson, J. B. *American Space: The Centennial Years, 1865–1876*. New York: W. W. Norton, 1972.

Jackson, Kenneth T. *Crabgrass Frontier: The Suburbanization of America*. New York: Oxford University Press, 1985.

Jaher, Frederic Cople. *The Urban Establishment: Upper Strata in Boston, New York, Charlestown, Chicago and Los Angeles*. Urbana: University of Illinois Press, 1982.

Kahrl, William L. *The California Water Atlas*. Sacramento: Governor's Office of Planning and Research, 1979.

———. *Water and Power: The Conflict over Los Angeles' Water Supply in the Owens Valley*. Los Angeles: University of California Press, 1982.

———. "Water for California Cities: Origins of the Major Systems." *Pacific Historian* 27 (1983): 17–23.

Kazin, Michael. *Barons of Labor: The San Francisco Building Trades and Union Power in the Progressive Era*. Urbana: University of Illinois, 1987.

Kelley, Robert L. *Battling the Inland Sea: American Political Culture, Public Policy, and the Sacramento Valley, 1850–1986*. Berkeley: University of California Press, 1989.

Kingdon, John W. *Agendas, Alternatives, and Public Policies*. Boston: Little, Brown, 1984.

Kinnicutt, Leonard Parker. "The Prevention of the Pollution of Streams by Modern Methods of Sewage Treatment." *Science* 14 (1902): 161–71.

Kolko, Gabriel. *The Triumph of Conservatism: A Reinterpretation of American History, 1900–1916*. Chicago: Quadrangle Books, 1967.

Koren, John. *Boston, 1822 to 1922: The Story of Its Government and Principal Activities During One Hundred Years*. Boston: 1923.

Leavitt, Judith Walzer. *The Healthiest City: Milwaukee and the Politics of Health Reform*. Princeton, N.J.: Princeton University Press, 1982.

Leavitt, Judith Walzer, and Numbers, Ronald L., eds. *Sickness and Health in America: Readings in the History of Medicine and Public Health*. Madison: University of Wisconsin Press, 1985.

Lerner, Barron H. "New York City's Tuberculosis Control Efforts: The Historical Limitation of the 'War on Consumption.'" *American Journal of Public Health* 83 (1993): 758–65.

Link, Vernon B. *A History of Plague in the United States of America.* Public Health Monographs No. 26. Washington, D.C.: 1955.

Little, Timothy G. "Court Appointed Special Masters in Complex Environmental Litigation: City of Quincy v. Metropolitan District Commission." *Harvard Environmental Law Review* 8 (1984): 435–75.

Lotchin, Roger. *San Francisco, 1846–1856: From Hamlet to City.* New York: Oxford University Press, 1974.

Luckin, Bill. *Pollution and Control: A Social History of the Thames in the Nineteenth Century.* Bristol, England: Hilger, 1986.

Mainwaring, W. F. B. Massey. *The Preservation of Fish Life in Rivers by the Exclusion of Town Sewage.* London: William Clowes & Sons, 1883.

Mann, Arthur. *Yankee Reformers in the Urban Age.* Cambridge, Mass.: Belknap Press, 1954.

Marcus, Alan I. "The City as Social System: The Importance of Ideas." *American Quarterly* 37 (1985): 332–45.

May, Judith V. "Progressive and the Poor: An Analytic History of Oakland," presented at Public Administration and Neighborhood Control conference, Boulder, Colorado, May 6–8, 1970.

McAfee, Ward M. *California's Railroad Era, 1850–1911.* San Marino, Calif.: Golden West Books, 1973.

———. "A Constitutional History of Railroad Rate Regulation in California, 1879–1911." *Pacific Historical Review* 37 (1968): 265–80.

———. "Local Interests and Railroad Regulation in Nineteenth Century California." Ph. D. dissertation, Stanford University, 1965.

———. "Local Interests and Railroad Regulation in California During the Granger Decade." *Pacific Historical Review* 37 (1968): 51–66.

McArdle, Phil, ed. *Exactly Opposite the Golden Gate.* Berkeley: Berkeley Historical Society, 1983.

McCaffrey, George. "The Disintegration and Reintegration of Metropolitan Boston." Ph.D. dissertation, Harvard University, 1937.

McCormick, Richard L., ed. *The Party Period and Public Policy: American Politics from the Age of Jackson to the Progressive Era.* New York: Oxford University Press, 1986.

McDonald, Terrence J. *The Parameters of Urban Fiscal Policy: Socioeconomic Change and Political Culture in San Francisco, 1860–1906.* Berkeley: University of California Press, 1986.

McEvoy, Arthur. *The Fishermen's Problem: Ecology and Law in the California Fisheries, 1850–1980.* Cambridge: Cambridge University Press, 1986.

McLean, Walter. *From Pardee to Buckhorn: Water Resources Engineering and Water Policy in the East Bay Municipal Utility District, 1927–1991.* Berkeley: University of California Oral History Office, 1993.

McMeiken, J. Elizabeth. *Public Health Professionals and the Environment: A Study of Perceptions and Attitudes.* Social Science Series, vol. 5. Ottawa: Inland Waters Directorate, Water Planning and Management Branch, 1972.

McNary, C. H. "Competition, Regulation or Municipal Ownership." *Pacific Municipalities* 26 (1912): 228–32.

Melosi, Martin V., ed. *Garbage in the Cities: Refuse, Reform and the Environment, 1880–1980.* College Station: Texas A&M University Press, 1981.

———. *Pollution and Reform in American Cities, 1870–1930.* Austin: University of Texas Press, 1980.

Melvin, Patricia Mooney. *The Organic City: Urban Definition and Community Organization, 1880–1920.* Lexington: University Press of Kentucky, 1987.

Moehring, Eugene P. *Public Works and the Patterns of Urban Real Estate Growth in Manhattan, 1835–1894.* New York: Arno Press, 1981.

Mohl, Raymond A. *The New City: Urban America in the Industrial Age, 1860–1920.* Arlington Heights, Ill.: Harlan Davidson, 1985.

Mohr, James. *The Radical Republicans and Reform in New York During Reconstruction.* Ithaca: Cornell University Press, 1973.

Monkkonen, Eric. *America Becomes Urban: The Development of United States Cities and Towns, 1780–1980.* Berkeley: University of California Press, 1988.

Mowry, George E. *The California Progressives.* Berkeley: University of California Press, 1951.

Mumford, Lewis. *The Brown Decades: A Study of Arts in America, 1865–1895.* New York: Dover, 1971.

Nash, Gerald D. "Bureaucracy and Economic Reform: The Experience of California, 1899–1911." *Western Political Quarterly* 13 (1960): 678–961.

———. "The California Railroad Commission, 1876–1911." *Southern California Quarterly* 44 (1962): 287–306.

———. "Stages in California's Economic Growth, 1870–1970: An Interpretation." *California Historical Quarterly* 51 (1972): 315–31.

———. *State Government and Economic Development: A History of Administrative Policies in California, 1849–1933.* University of California, Berkeley: 1964.

Nash, Roderick. *Wilderness and the American Mind.* New Haven, Conn.: Yale University Press, 1973.

Nelson, William. *Roots of American Bureaucracy: 1830–1900.* Cambridge, Mass.: Harvard University Press, 1982.

Nesson, Fern L. *Great Waters: A History of Boston's Water Supply.* Hanover, N.H.: University Press of New England, 1983.

Noble, John Wesley. *Its Name Was M.U.D.* Oakland, Calif.: East Bay Municipal Utility District, 1970.

Norman, Robert Toll. "The Metropolitan State: The Urbanization of Greater Boston." Ph.D. dissertation, Harvard University, 1963.

O'Connor, Thomas H. *The Boston Irish: A Political History.* Boston: Northeastern University Press, 1995.

Odiorne, James Creighton. *Genealogy of the Odiorne Family.* Boston: Rand, Avery, 1875.

Ogle, Maureen. *All the Modern Conveniences: American Household Plumbing, 1840–1890.* Baltimore: Johns Hopkins University Press, 1996.

Olin, Spencer, C., Jr. *California Politics, 1846–1920: The Emerging Corporate State.* San Francisco: Boyde & Fraser, 1981.

———. *California's Prodigal Sons: Hiram Johnson and the Progressives, 1911–1917.* Berkeley: University of California Press, 1968.

Orsi, Richard J. "The Octopus Reconsidered: The Southern Pacific and Agricultural Modernization in California, 1865–1915." *California Historical Quarterly* 54 (1975): 197–220.

______. "Railroads and Water in the Arid Far West: The Southern Pacific Company as a Pioneer Water Developer." *California History* 70 (1991): 46–61.

Ostrom, Elinor. "The Social-Stratification-Government Inequality Thesis Explored." *Urban Affairs Quarterly* 17 (1983): 91–112.

Owens, John R.; Costantini, Edmond; and Weschler, Louis F. *California: Politics and Parties.* New York: Macmillan, 1970.

Palmer, Tim. *Endangered Rivers and the Conservation Movement.* Berkeley: University of California Press, 1986.

Pernick, Martin. "Politics, Parties and Pestilence." In Judith Walzer Leavitt and Ronald L. Numbers, eds., *Sickness and Health in America: Readings in the History of Medicine and Public Health* (Madison: University of Wisconsin Press, 1985), pp. 356–71.

Peterson, Jon A. "Environment and Technology in the Great City Era." *Journal of Urban History* 8 (1982): 343–54.

______. "Impact of Sanitary Reform upon American Urban Planning, 1840–1890." *Journal of Social History* 13 (1979): 83–104.

Phillips, David J. H. *Toxic Contaminants in the San Francisco Bay-Delta and Their Possible Biological Effects.* Richmond, Calif.: Aquatic Habitat Institute, 1986.

Pisani, Donald J. "Enterprise and Equity: A Critique of Western Water Law in the Nineteenth Century." *Western Historical Quarterly* 18 (1987): 15–37.

______. *From the Family Farm to Agribusiness: The Irrigation Crusade in California and the West, 1850–1931.* Berkeley: University of California Press, 1984.

______. "Promotion and Regulation: Constitutionalism and the American Economy." *Journal of American History* 74 (1987): 740–68.

______. *To Reclaim a Divided West: Water, Law and Public Policy, 1848–1902.* Albuquerque: University of New Mexico Press, 1992.

Platt, Howard L. *City Building in the New South: The Growth of Public Services in Houston, TX, 1830–1910.* Philadelphia: Temple University Press, 1983.

Posner, Russell M. "The Progressive Voters League, 1923–1926." *California Historical Quarterly* 36 (1957): 251–62.

Public Health in the Victorian Age: Debates on the Issue from Nineteenth Century Critical Journals. Westmead, England: Gregg International, 1973.

Public Health Reports and Papers Presented at the Meetings of the American Public Health Association in the Years 1874–1875. New York: Hurd & Houghton, 1876.

Putnam, Jackson K. "Persistance of Progressivism in the 1920s: The Case of California." *Pacific Historical Review* 35 (1966): 395–412.

Rather, Lois. *Oakland's Image: A History of Oakland, California.* Oakland: Rather Press, 1972.

Reid, Donald. *Paris Sewers and Sewermen: Realities and Representations.* Cambridge, Mass.: Harvard University Press, 1991.

Reisner, Marc. *Cadillac Desert: The American West and Its Disappearing Water.* New York: Viking Press, 1986.

Reisner, Marc, and Bates, Sarah. *Overtapped Oasis: Reform or Revolution for Western Water.* Washington, D.C.: Island Press, 1990.

Reps, John W. *Cities of the American West: A History of Frontier Urban Planning.* Princeton, N.J.: Princeton University Press, 1979.

Rice, Bradley R. *Progressive Cities: The Commission Government Movement in America, 1901–1920.* Austin: University of Texas Press, 1977.

River, Robert B. *Efficiency, Responsibility and Accomplishment of the East Bay Municipal Utility District.* Oakland: Institute for Public Utility Research, 1954.

Rogin, Michael Paul, and Shover, John L. *Political Change in California: Critical Elections and Social Movements, 1890–1966. Contributions to American History,* vol. 5., ed. by Stanley I. Kutler. Westport, Conn.: Greenwood, 1970.

Rosen, Christine Meisner. *The Limits of Power: Great Fires and the Process of City Growth in America.* Cambridge: Cambridge University Press, 1986.

Rosen, Howard, and Keating, Ann, eds. *Water and the City: The Next Century.* Chicago: Public Works Historical Society, 1991.

Rosenberg, Charles. *The Cholera Years: The United States in 1832, 1849 and 1866.* Chicago: University of Chicago Press, 1962.

Rosenkrantz, Barbara G. *Public Health and the State: Changing Views in Massachusetts, 1842–1936.* Cambridge, Mass.: Harvard University Press, 1972.

Russell, James Michael. *Atlanta, 1847–1890: City Building in the Old South and the New.* Baton Rouge: Louisiana State University Press, 1988.

Russell, Janet N., and Berryman, Jack W. "Parks, Boulevards and Outdoor Recreation: The Promotion of Seattle." *Journal of the West* 26 (1987): 5–17.

Ryan, Dennis P. *Beyond the Ballot Box: A Social History of the Boston Irish, 1845–1917.* Rutherford, N.Y.: Fairleigh Dickinson University Press, 1983.

Saxton, Alexander. *Indispensible Enemy: Labor and the Anti-Chinese Movement in California.* Berkeley: University of California Press, 1971.

Schiesl, Martin J. *The Politics of Efficiency: Municipal Administration and Reform in America, 1800–1920.* Berkeley: University of California Press, 1977.

Schmeckerbier, Laurence F. *The Public Health Service; Its History, Activities and Organization.* Baltimore: Johns Hopkins Press, 1923.

Schmitt, Peter J. *Back to Nature:The Arcadian Myth in Urban America.* New York: Oxford University Press, 1969.

Schultz, Stanley K. *Constructing an Urban Culture: American Cities and City Planning.* Philadelphia: Temple University Press, 1989.

Schultz, Stanley K., and McShane, Clay. "To Engineer the Metropolis: Sewers, Sanitation and City Planning in Late-Nineteenth Century America." *Journal of American History* 65(2) (1978): 389–411.

Schuyler, David. *The New Urban Landscape: Redefinition of City Form in Nineteenth Century America.* Baltimore: Johns Hopkins University Press, 1986.

Scott, Mel. *American City Planning Since 1890; A History Commemorating the Fiftieth Anniversary of the American Institute of Planners.* Berkeley: University of California Press, 1969.

———. *San Francisco Bay Area: A Metropolis in Perspective,* 2d ed. Berkeley: University of California Press, 1985.

Scott, Stanley, and Bollens, John C. *Special Districts in California Local Government. Legislative Problems,* vol. 4. Berkeley: Bureau of Public Administration, 1949.

Scott, Thomas. "The Diffusion of Urban Government Forms as a Case of Social Learning." *Journal of Politics* 30 (1968): 1091–1108.

Simonds, Thomas C. *History of South Boston: Formerly Dorchester Neck, New Ward XII of the City of Boston*. Boston: David Clapp, 1857.

Sitton, Tom. "John Randolph Haynes and the Left Wing of California Progressivism." In William Deverell and Tom Sitton, eds., *California Progressivism Revisited* (Berkeley: University of California Press, 1944).

Skocpol, Theda. *Protecting Soldiers and Mothers: The Political Origins of Social Policy in the United States*. Cambridge, Mass.: Belknap Press, 1992.

Skocpol, Theda, and Ikenberry, John. "The Political Formation of the American Welfare State in Historical and Comparative Perspective." *Comparative Social Research* 6 (1983): 87–148.

Skowronek, Stephen. *Building a New American State: The Expansion of National Administrative Capacities, 1877–1920*. Cambridge: Cambridge University Press, 1982.

Slater, Kelly. "The Port of Boston." *Sanctuary* (July/August 1986): 3–7.

Smillie, Wilson G. *Public Health: Its Promise for the Future: A Chronicle of the Development of Public Health in the United States, 1607–1914*. New York: Macmillan, 1955.

Smith, Robert G. *Ad Hoc Governments: Special Transportation Authorities in Britain and the United States*. Beverly Hills, Calif.: Sage Publications, 1974.

Smith, Robert L., ed. *The Ecology of Man: An Ecosystem Approach*. New York: Harper & Row, 1972.

Snyder, James Barton. "Floods upon Dry Ground: A History of Water Law and Water Resource Development in California, 1900–1928." Ph.D. dissertation, University of California at Davis, 1967.

Sofen, Edward. *The Miami Metropolitan Experiment*. Bloomington: Indiana University Press, 1963.

Spangler, Frank L. *Operation of Debt and Tax Rate Limits in the State of New York*. Albany, N.Y.: J. B. Lyon, 1932.

Steinberg, Theodore. *Nature Incorporated: Industrialization and the Waters of New England*. Cambridge: Cambridge University Press, 1991.

Stilgoe, John R. *Borderland: Origins of the American Suburb 1820–1939*. New Haven, Conn.: Yale University Press, 1989.

Sweeney, Stephen B., and Blair, George S., eds. *Metropolitan Analysis: Important Elements of Study and Action*. Philadelphia: University of Pennsylvania Press, 1958.

Tarr, Joel. *The Search for the Ultimate Sink: Urban Pollution in Historical Perspective*. Akron, Ohio: University of Akron Press, 1996.

______. "The Separate versus Combined Sewer Problems: A Case Study in Urban Technology Design Choice." *Journal of Urban History* 5 (1979): 308–40.

Tarr, Joel, and Dupuy, Gabriel, eds. *Technology and the Rise of the Networked City in Europe and America*. Philadelphia: Temple University Press, 1988.

Tarr, Joel A.; McCurley, James; and Yosie, Terry F. "The Development and Impact of Urban Wastewater Technology: Changing Concepts of Water Quality Control, 1850–1930." In Martin V. Melosi, ed., *Pollution and Reform in American Cities, 1870–1930* (Austin: University of Texas Press, 1980), pp. 62–78.

Teaford, Jon. *City and Suburb: The Political Fragmentation of Metropolitan America, 1850–1970*. Baltimore: Johns Hopkins University Press, 1979.

______. *The Municipal Revolution in America: Origins of Modern Urban Government, 1650–1825*. Chicago: University of Chicago Press, 1975.

———. *The Twentieth-Century American City: Problem, Promise, and Reality.* Baltimore: Johns Hopkins University Press, 1986.

———. *Unheralded Triumph: City Government in America, 1870–1900.* Baltimore: Johns Hopkins University Press, 1984.

Thelan, Max. "The Public Utilities Act and Its Relation to Municipalities." *Pacific Municipalities* 26 (1912): 49–55.

Thomas, June Manning. "Racial Crisis and the Fall of the Detroit City Plan Commission." *Journal of the American Planning Association* 54 (1988): 150–61.

Titus, Charles H. "Voting in California Cities, 1900–1925." *Southwestern Social Science Quarterly* 8 (1928): 381–99.

Tomes, Nancy. "The Private Side of Public Health: Sanitary Science, Domestic Hygiene, and the Germ Theory, 1870–1900." *Bulletin of the History of Medicine* 64 (1990): 509–39.

Van Tassel, David D., and Grabowski, John J., eds. *Cleveland: A Tradition of Reform.* Kent, Ohio: Kent State University Press, 1986.

Wain, Henry. *A History of Preventive Medicine.* Springfield, Ill.: Charles C. Thomas, 1970.

Walker, Richard A., and Williams, Matthew J. "Water from Power: Water Supply and Regional Growth in the Santa Clara Valley." *Economic Geography* 58 (1982): 95–119.

Walsh, Annmarie Hauck. *The Public's Business: The Politics and Practices of Government Corporations.* Cambridge, Mass.: MIT Press, 1978.

Walton, John. *Western Times and Water Wars: State, Culture and Rebellion in California.* Berkeley: University of California Press, 1992.

Warner, Margaret. "Local Control versus National Interest: The Debate over Southern Public Health, 1878–1884." *Journal of Southern History* 50 (1984): 407–28.

Warner, Sam Bass, Jr. *The Private City: Philadelphia in Three Periods of Its Growth.* Philadelphia: University of Pennsylvania Press, 1968.

———. *Streetcar Suburbs: The Process of Growth in Boston, 1870–1900.* Cambridge, Mass.: Harvard University Press, 1962.

———. *Urban Wilderness: A History of the American City.* New York: Harper & Row, 1972.

Warren, Robert O. *Governments in Metropolitan Regions: A Reappraisal of Fractionated Political Organization.* Davis, Calif.: Institute of Governmental Affairs, 1966.

Watson, Richard A., and Romani, John H. "Metropolitan Government for Metropolitan Cleveland: An Analysis of Voting Records." *Midwest Journal of Political Science* 5 (1961): 365–90.

Weidner, Charles H. *Water for a City: A History of New York City's Problem from the Beginning to the Delaware River System.* New Brunswick, N.J.: Rutgers University Press, 1974.

Whipple, George C. *State Sanitation: A Review of the Work of the Massachusetts State Board of Health.* New York: Arno Press, 1877.

Willard, Ruth Hendricks. *Alameda County, California Crossroads: An Illustrated History.* Windsor Publications, 1988.

Wilson, James Q., and Banfield, Edward C. "Voting Behavior on Municipal Public Expenditures: A Study in Rationality and Self-Interest." In Julius Margolis, ed.,

The Public Economy of Urban Communities (Washington, D.C.: Resources for the Future, 1965), pp. 74–91.

Wilson, William H. *The City Beautiful Movement.* Baltimore: Johns Hopkins University Press, 1989.

Worster, Donald, ed. *The Ends of the Earth: Perspectives on Modern Environmental History.* Cambridge: Cambridge University Press, 1988.

______. *Rivers of Empire: Water, Aridity and the Growth of the American West.* New York: Pantheon Books, 1985.

Zimmerman, Joseph F., ed. *Government of the Metropolis.* New York: Holt, Rinehart & Winston, 1968.

Zink, Harold. *City Bosses in the United States: A Study of Twenty Municipal Bosses.* Durham: Duke University Press, 1930.

Index